D0759722

# ANALYTICAL SOLID-PHASE EXTRACTION

# ANALYTICAL SOLID-PHASE EXTRACTION

**JAMES S. FRITZ**
Department of Chemistry and Ames Laboratory
Iowa State University, Ames, IA 50011

New York / Chichester / Weinheim / Brisbane / Singapore / Toronto

This book is printed on acid-free paper.



Published simultaneously in Canada.



For ordering and customer service, call 1-800-CALL-WILEY.

***Library of Congress Catalog Card Number: 67-13943***

Fritz, James S. (James Sherwood), 1924-
Analytical solid-phase extraction / James S. Fritz.
p. cm.
Includes bibliographical references and index.
ISBN 0-471-24667-0 (alk. paper)
1. Extraction (Chemistry) I. Title
QD63.E88F68 1999
543'.0892--dc21 98-43758

Printed in the United States of America.

10 9 8 7 6 5 4 3 2

Dedicated to my wife, Mim

# CONTENTS

# PREFACE

In chemistry, as in other endeavors, the simplest ideas are often the best. Extraction of solutes from an aqueous analytical sample by a liquid organic solvent has been in use for nearly a century (at the time of writing). But organic solvents pollute the air and water and can be messy to work with. So why not use small particles of a porous organic polymer as an extractant in place of a liquid solvent?

The feasibility of this approach was confirmed in the early 1970s; numerous publications followed, but it has taken some additional years for its value to be fully appreciated. A symposium at the 1998 Pittsburgh Conference was entitled, "Solid-phase extraction—a neglected orphan." But this symposium was really a celebration of the fact that solid-phase extraction (SPE) is no longer a neglected technique. On the contrary, SPE is undergoing a remarkable period of growth and increasing popularity.

This book was written for scientists who want to know more about SPE for analytical use. All major aspects of SPE are covered: basic principles, historical perspective, materials and equipment, extraction of organic solutes from aqueous samples, extraction of polar solutes from apolar organic solutions, ion-exchange SPE, extraction of metal ions, use of membrane disks, solid-phase extraction on microscale and semimicroscale, and selected applications. The goal is to provide the reader with a real understanding of how SPE works (principles and theory) as well as what one can do with SPE (scope and limitations). This knowledge will be useful in making intelligent choices among published SPE procedures and in modifying conditions to meet new situations. It is also hoped that researchers will be stimulated to develop entirely new and better methods for SPE.

JAMES S. FRITZ

*Ames, Iowa*

# ACKNOWLEDGMENTS

This book is to a considerable extent the result of the author's own experience in the field of solid-phase extraction. Gaining this experience would not have been possible without the dedicated research of numerous students and other scientific collaborators. The following persons played an important role in the early development of solid-phase extraction: Mike Arguello, Mike Avery, Vince Calder, Richard Chang, Colin Chriswell, Bonnie Glatz, Gregor Junk, Larry Kissinger, John Richard, Nancy Nearing, Harry Svec, Akira Tateda, Ray Vick, Ray Willis, David Witiak, and Joanne Witiak.

More recent colleagues have also contributed greatly to the development of SPE: SPE of metal ions—Intessar Albiaty, Ron Freeze, Jeff King, and Michelle Thornton; development of new resins—Michael Büchmeiser, Phil Dumont, Doug Gjerde, Don Hagen, Luther Schmidt, and Jeff Sun; and special techniques—Dianna (Mayer) Ambrose, Tom Chambers, Jeremy Masso, and Luther Schmidt.

Special thanks go to Craig Markell, who read the entire manuscript and made many helpful comments; to Jeremy Masso for providing numerous literature references; to Marilyn Kniss, who carefully and patiently typed the manuscript with its several revisions; and to Cathy Hertz for an excellent copyediting job. The author also wishes to thank Tom Chambers, Günter Grienberger, Craig Markell, Luther Schmidt, and Jeff Sun for providing some of the illustrations used in the book.

James S. Fritz

*Ames, Iowa*

# CHAPTER 1

# INTRODUCTION AND PRINCIPLES

## 1.1 INTRODUCTION

### 1.1.1 The Importance of Sample Preparation

Analytical laboratories are under great pressure to provide analyses more quickly and at lower cost. This burden falls most heavily on the sample preparation portion of the laboratory, which is asked to provide more reproducible results, accommodate lower technical skills, decrease the use of organic solvents, provide cleaner extracts for instrumental measurement, and do everything more quickly and at less cost.

The reason why there is such a need for improvement in sample preparation techniques is that the majority of the sample analysis time is spent in preparing the sample. One study (1) showed that more than 60% of analysis time was spent in sample preparation compared to only about 7% for the actual measurement of the sample constituents. The remainder of the time was taken up by sample collection and data handling.

In the past liquid–liquid extraction has played a major role in sample cleanup and concentration of the sample components to be measured. However, recovery of sample components by liquid extraction is seldom complete. Liquid extraction tends to be slow and labor-intensive. More stringent environmental concerns are making the use and disposal of large amounts of organic solvents more difficult.

The popularity and use of solid-phase extraction (SPE) is growing at a fast rate. SPE is easily automated, faster, and in general more efficient than liquid–liquid extraction. The particles used in SPE are nonpolluting and the amount of liquid solvents used are tremendously lower than in liquid–liquid extraction.

### 1.1.2 What is Solid-Phase Extraction?

In solid-phase extraction solutes are extracted from a liquid phase into a solid phase. The solid phase typically consists of small, porous particles of silica with a bonded organic phase or of an organic polymer, such as crosslinked polystyrene. The extraction can take place in a batch mode in which the solid extractant is intimately mixed with the liquid sample solution. In chemical analysis it is more common to pack the solid extractant into a small tube and pass the liquid sample through the tube. A typical apparatus for SPE is shown in Figure 1.1.

Solid-phase extraction is not limited to the use of solid particles to extract solutes from a liquid sample. Air or other gaseous samples can also be passed through a packed tube to extract organic vapors or other substances present in the sample.

Substances that have been extracted by the solid particles can be removed by washing with an appropriate liquid solvent. For example, most organic analytes can be eluted from a SPE tube (column) with an organic solvent such as acetone, acetonitrile, or methanol. Usually, the volume of solvent needed for complete elution of the analytes is much smaller than the original sample volume. A concentration of the analytes is thus achieved.

Extracted molecules can often be removed from solid particles by heating in a gentle stream of a nonreactive carrier gas. This can be a convenient way to transfer the molecules into a gas chromatograph for analysis.

### 1.1.3 History and Literature

The history of solid-phase extraction dates back at least to the early 1970s, when columns packed with Rohm and Haas XAD resin particles were used to concentrate very low concentrations of organic pollutants from water samples (2). However, activated carbon had been used for several years prior to 1970 to accumulate organic solutes prior to analysis. The earlier work on SPE is discussed in some detail in Chapter 2.

During the late 1980s and the 1990s, the development and use of analytical solid-phase extraction have expanded tremendously. A number

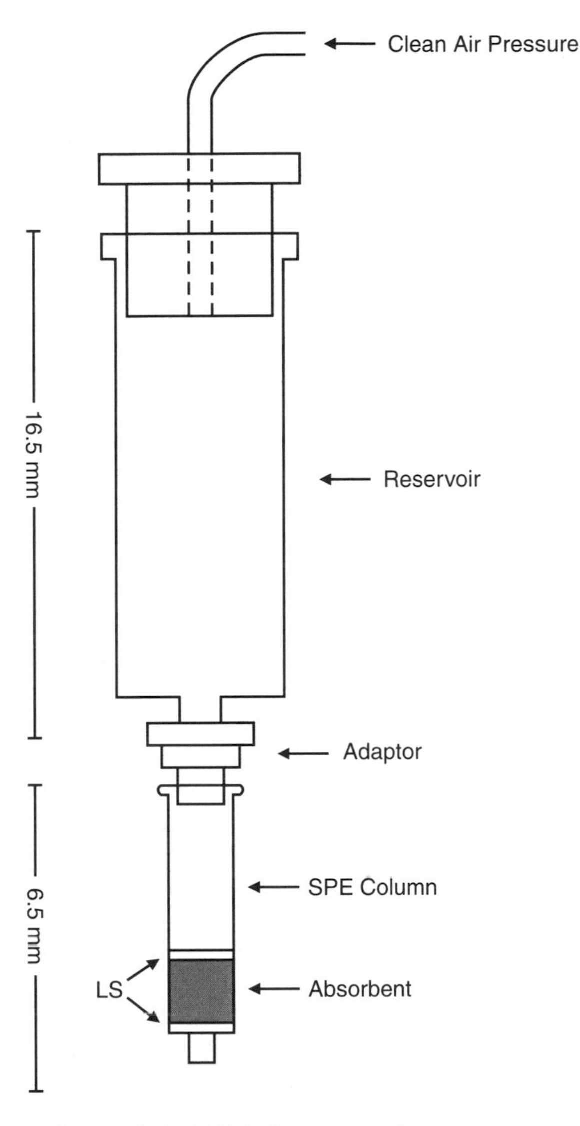


**Figure 1.1** Solid-phase extraction apparatus.

of reviews and symposia on SPE have been published (3–5). In addition, at least two books on SPE have been published (6,7). At this writing, several companies that sell equipment and supplies for SPE offer booklets on its use and extensive bibliographies of SPE applications. These include the following: J. T. Baker, Inc., Phillipsburg, NJ; 3M Co., St. Paul, MN; Varian, Harbor City, CA; and Waters, Milford, MA.

## 1.2 PRINCIPLES

### 1.2.1 Comparison of SPE with Liquid–Liquid Extraction

In liquid–liquid extraction the objective is to transfer the desired solutes from one liquid solution to another, nonmiscible liquid. Most commonly, the solutes are extracted from a predominately aqueous solution into an organic liquid. Analytical liquid–liquid extraction is usually performed in a separatory funnel so that the two liquid phases can be separated after the extraction. The extractive liquid may be either heavier than water or lighter than water.

To carry out an extraction, the extractive solvent is added to the aqueous sample solution, the vessel is tightly stoppered, and the vessel is shaken vigorously to create a temporary emulsion. The emulsion consists of very small spherical droplets of the extractive liquid suspended in the aqueous phase. The interfacial contact area between the two phases must be quite large in order to promote rapid mass transfer of the desired solutes from one phase to another. Even so, it is sometimes necessary to continue shaking for several minutes to attain a true equilibrium of the solutes between the two phases. When shaking is terminated, the emulsion should break up and the two liquid coalesce to form two continuous but nonmiscible liquid phases. A practical difficulty of liquid–liquid extraction is that emulsions sometimes break up very slowly or incompletely.

In classic analytical liquid extraction, when there is no longer an emulsion the bottom liquid is carefully drained from the separatory funnel by opening the stopcock very carefully until the last of the lower liquid has just been drained off. This can be a tedious operation. When chemical analysis of the extracted solutes cannot be carried out directly in the organic liquid, it is necessary to do a backextraction. This can often be accomplished with an aqueous solution under conditions (e.g., changed pH) causing the solutes to prefer the aqueous phase.

Some of the concepts of liquid–liquid extraction apply to solid-phase extraction. Instead of a temporary liquid emulsion, the extractive material in SPE is a suspension of spherical solid particles in the aqueous sample. The extractive particles typically are ~10–50 μm in diameter with a very large surface area, often ~200–800 $m^2/g$. A large interfacial area between the particles and the sample solution is needed for rapid mass transfer of the extracted solutes from one phase to the other. The extractive particles must be sufficiently dense and large enough in diameter to settle rapidly when agitation of the solid and liquid phases is terminated.

Although SPE can be done in a batch equilibration similar to that used in liquid–liquid extraction, it is much more common to use a small tube (minicolumn) or cartridge packed with the solid particles. The liquid sample is passed through the column, thus coming into intimate contact with the solid particles. With modern solid extractants, equilibrium is rapidly attained and the analytes tend to be extracted in a zone near the top of the SPE column. Unlike batch extraction, where there is only a single equilibration of solutes between the two phases, there are in effect multiple equilibrations when SPE is performed with a packed minicolumn. This is because the solutes continuously encounter fresh particles (containing little, if any, extracted solutes) as they pass down the column. A higher percentage of extraction is thus expected in column SPE compared to batch-type extractions.

It is usually necessary to transfer extracted analytes from the solid particles to a liquid phase for final measurement. The chemistry of this step will depend on the types of analytes and solid extractants that are used. Organic compounds can usually be eluted from a SPE minicolumn by a small volume ($<1$ ml) of a liquid organic solvent.

### 1.2.2 Comparison Between SPE and HPLC

Some confusion exists concerning the distinction between solid-phase extraction and liquid chromatography. Both use a small tube, or column, packed with a porous solid that acts as the stationary phase. Multiple equilibria for the various analytes exist between the liquid and solid phases. These equilibria can perhaps be best expressed in terms of the capacity factors (also called *retention factor*), $k$, of the various sample components (analytes). For any given substance the capacity factor is defined as

$$k = \frac{\text{amount in the solid phase}}{\text{amount in the liquid phase}} \quad (1.1)$$

When the liquid phase is flowing through the packed column at a linear flow rate of $u$ cm/s, the rate of movement of an analyte band through the column is $u/(1 + k)$. When $k$ is quite large (e.g., 100), there is almost no movement of the analyte band. As $k$ decreases, the rate of movement of an analyte band becomes faster and approaches that of the liquid itself ($u$).

The capacity factor of an analyte usually will vary over a broad range, depending on the composition of the liquid phase. In liquid chromatography the most widely used liquid phases are a mixture of water and an organic solvent. The capacity factor of an analyte becomes progressively smaller as the organic content of the liquid phase is increased and the analyte becomes more strongly solvated by the organic liquid. Typical plots of $k$ and log $k$ against $\phi$, the volume fraction of organic solvent in the liquid phase, are given in Figures 1.2 and 1.3. The plot of log $k$ versus $\phi$ is quadratic, but is often approximately linear over the middle range of $\phi$.

In liquid chromatography the intent is to separate various sample components from one another on the basis of different rates of migration through the column. The mobile-phase composition is adjusted so that the sample components will have capacity factors (also called *retention factors*) between about 1 and 10. A very small volume of sample (e.g., 5–50 μl) is injected into the top of the column while the mobile liquid phase is continuously pumped through the column. The analytes are separated by virtue of their different migration rates and are detected by an appropriate detector at the end of the column.

In SPE the sample solution itself is the mobile phase. Ideally, the capacity factors of the sample components will be very high (100, 1000, or more). Under these conditions the sample components will be retained as a single, tight band on the column. Since the column typically is short, the band of

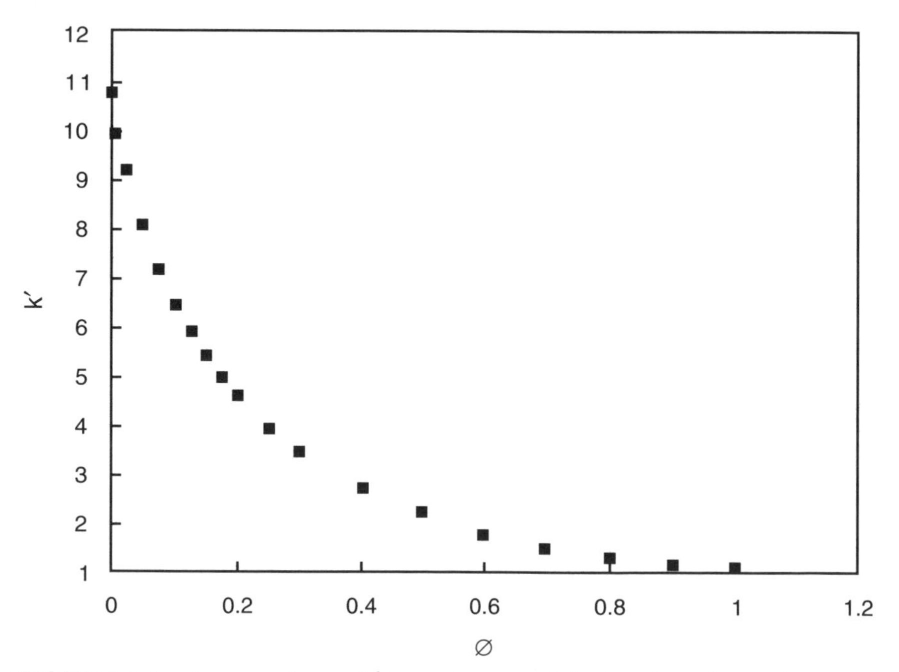


**FIGURE 1.2** Capacity factor ($k'$) as a function of the fraction of organic solvent ($\phi$).

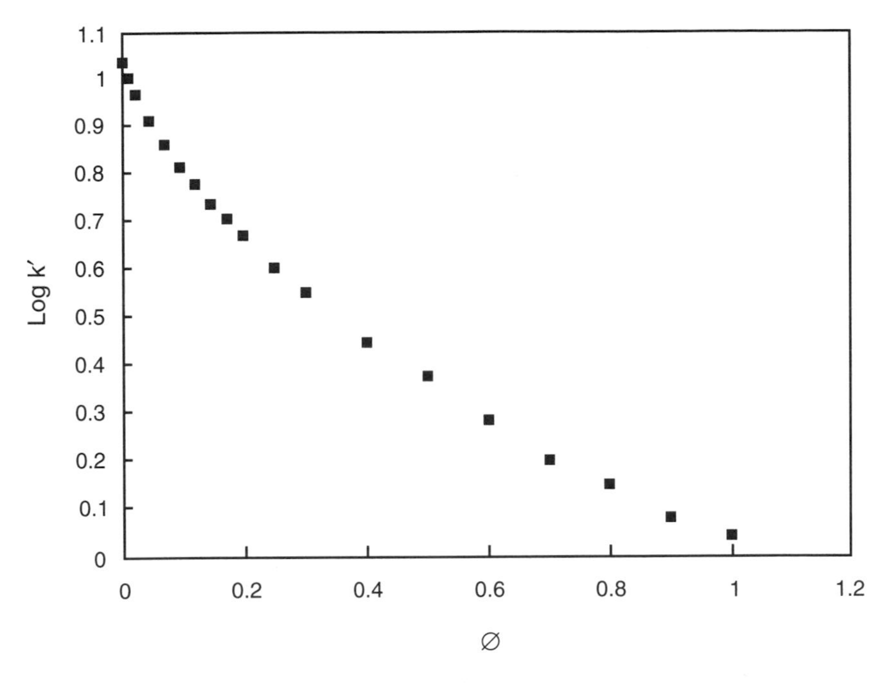


**FIGURE 1.3** Log $k'$ as a function of $\phi$.

extracted material may occupy 50% or more of the column length. After a brief rinse (usually with water), the extracted materials are eluted rapidly as a group under conditions where the capacity factors will be as low as possible. For example, organic compounds can usually be eluted by < 1 ml of an organic solvent. Measurement of the individual sample components is usually accomplished by some form of chromatography on a portion of the eluate from the SPE column.

There is ordinarily no separation of individual analytes in SPE. All the analytes are retained as a sharp band on the solid phase and then are eluted together by a liquid organic solvent. However, it sometimes is possible to separate the analytes into two or more groups by judicious choice of the solid phase and eluting conditions. For example, both neutral compounds and protonated organic bases are retained on a porous resin containing a fairly low concentration of sulfonate groups. The neutral compounds are then eluted by an organic solvent, but the protonated bases are only eluted by an organic solvent containing a base to remove the protons from analyte bases.

The differences between liquid extraction, solid-phase extraction, and liquid chromatography are summarized concisely in Table 1.1.

**TABLE 1.1 Comparison Between Liquid–Liquid Extraction, SPE, and Liquid Chromatography (LC)**

| | Liquid Extraction | SPE | LC |
|---|---|---|---|
| Extractive phase | Liquid emulsion | Porous solid | Porous solid |
| Typical sample | $H_2O$, ≥ 10–100 ml | $H_2O$; ≥ 10–100 ml | $H_2O$–organic liquid or organic 5–50 μl |
| Partition equilibrium | Single | Multiple | Multiple |
| Separation of individual analytes | No | No | Yes |
| Easily automated | No | Yes | No |
| Elution from extractive phase | Seldom needed; $H_2O$, pH control | Organic liquid | $H_2O$–organic liquid mixture |
| Concentration achieved | Moderate | High | Slight |

## 1.2.3 Completeness of Extraction

Partitioning of a chemical solute between two phases follows the laws of chemical equilibrium. If a solute, A, exists in only a single chemical form, the distribution coefficient between two phases ($K_d$) is given by

$$K_d = \frac{[A]_2}{[A]_1} \tag{1.2}$$

where $[A]_2$ and $[A]_1$ are the concentrations of solute at equilibrium in the two phases. Usually phase 2 is organic and phase 1 is aqueous. The concentration of A in each phase is the mass (the amount) of A divided by the volume of that phase

$$K_d = \frac{(\text{Mass of A})_2 / V_2}{(\text{Mass of A})_1 / V_1} \tag{1.3}$$

The ratio of the mass of A in the two phases is defined as the *mass distribution ratio*, *D*. Substituting into Equation 1–3 (1.3), we obtain

$$K_d = \frac{D \cdot V_1}{V_2} \tag{1.4}$$

where $V_1$ and $V_2$ are the volumes of the two phases. Note that $K_d$ will be a constant for each system (a particular solute, A, in phases 1 and 2) whereas the value of $D_m$ will vary according to the volume ratio:

$$D = \frac{K_d \cdot V_2}{V_1} \tag{1.5}$$

where it will be seen that the extraction of a solute from phase 1 to phase 2 will be greater when the relative volume of phase 2 to phase 1 is increased. The fraction of total solute remaining in phase 1 after an extraction ($f$) is given by

$$f = \frac{1}{D+1} \tag{1.6}$$

The fraction extracted ($f_{ex}$) is given by

$$f_{ex} = \frac{D}{D+1} \tag{1.7}$$

If a series of $n$ extractions is performed in which the phases are separated after each extraction and the same volume as before of fresh phase 2 is added, the cumulative fraction of A not extracted is

$$f = \frac{1}{(D+1)^n} \tag{1.8}$$

The total fraction of A extracted from phase 1 into phase 2, is of course, equal to $(1 - f)$ and the percentage extraction is

$$\% \text{ extraction} = 100\,(1 - f) \tag{1.9}$$

In liquid–liquid extraction, phase 1 is normally water and phase 2 is a liquid organic solvent. However, the same relationships apply equally well to the situation where phase 1 is water or a water–organic solvent mixture, and phase 2 consists of solid extractive particles of the type used in solid-phase extraction. Indeed, liquid–liquid extraction and liquid–solid extraction (SPE) are similar techniques when each is carried out on a batch basis. When agitated vigorously, two immiscible liquid phases form a temporary emulsion with a large interfacial surface area for rapid mass transfer of solutes from one phase to the other. When agitation is stopped, the emulsion breaks and two distinct liquid phases are re-formed.

When SPE is performed on a batch basis, small, porous particles of solid extractant are added to the liquid sample. The temporary emulsion formed in liquid–liquid extraction and the porous nature of the particles in solid-phase extraction both provide the large interfacial surface area needed for efficient mass transfer of the solutes. Some agitation is helpful in speeding up the extraction step when SPE is carried out on a batch basis. When equilibrium has been attained between the two phases, the solid particles can be separated by filtration or simply by carefully pouring off the liquid phase.

Although SPE is feasible in a batch-equilibration mode, it is advantageous to use a short column packed with the solid extractant. Passing the liquid sample through such a column provides intimate contact with the solid particles, and there are multiple equilibria as the sample passes through the column and comes into contact with fresh solid extractant particles. In this mode the completeness of extraction will depend on the number of separate equilibrations with fresh solid extractant in the column as well as the mass distribution ratio. (In column operations the number of equilibrations is called the *theoretical plate number*.)

In a single equilibrium method such as simple liquid–liquid extraction, an "all or nothing" extraction is needed for a quantitative separation of a target solute from other substances in the sample. This means that extraction of the desired sample component should be nearly 100% while extraction of other components should be essentially zero. This situation is often difficult to achieve. SPE in a column provides multiple equilibrations and therefore requires only a reasonable *difference* in extractability to separate two solutes.

The separation of two substances, A and B, by solid-phase extraction can be illustrated by an example. Suppose that there are eight separate equilibrations with fresh resin in a SPE minicolumn. This is the same as saying that the column has eight theoretical plates. The fraction of a solute not extracted can be estimated from Equation (1.8).

To apply this equation to the SPE column, we will assume that the column contains equal volumes of solid-phase particles and aqueous sample. A solute, A, that is only 60% extracted by the first plate ($D$ =60/40 = 1.50) will be quantitatively extracted after being in contract with all eight plates:

$$f = \frac{1}{(1.5+1)^8} = 0.0007\ (0.07\%) \qquad \text{(extraction is 99.93\% complete)}$$

A solute, B, that is only 1% extracted by the first plate will have $D$ = 1/909 or approximately 0.01. After it has contracted all eight plates, then

$$f = \frac{1}{(1.01)^8} = 0.92\ (92\%) \qquad \text{(extraction is 8\% complete)}$$

Because of its very low $D$ value, this 8% of solute B can be quickly rinsed from the column by a little water, while the A, which is mostly near the top of the column, will hardly move. In this way a quantitative separation of A and B can be obtained.

### 1.2.4 Sample Breakthrough

When concentrating dilute samples by solid-phase extraction with a short column, the capacity of the column should be well in excess of the total analyte content of the sample. This being the case, the next important question concerns the maximum sample volume that can be passed through the SPE column before breakthrough occurs. This will depend on the retention volume of the analyte and the number of theoretical plates in the column.

Applications involving concentration by SPE differ from ordinary analytical chromatography in several respects; For instance, the sample enters the column as a front instead of a narrow plug. Since no separation of the analytes is intended, it is not necessary to have a large number of theoretical plates. Finally, the relative retention volume ($V_R/V_o$) must be large in order to permit the use of large sample volumes.

When a sample is fed continuously into a column, it is usually assumed that the shape of the eluting front (the breakthrough curve) can be described as the integral of a Gaussian peak. Lövkist and Jönsson (8) examined five equations that have been proposed to describe the shape of the eluting front. Although the results differed for columns containing only one or two theoretical plates, the profiles calculated from the different equations were in reasonable agreement when the column plate number ($n$) was 5 and in good agreement when $n = 25$. In Figure 1.4 the normalized flux $J/J_0$ (where $J_0$ and $J$ are the concentrations of analyte in the entering and eluting samples, respectively) is plotted against the relative retention, $\tau$ ($T = V/V_R$).

In calculating the plots in Figure 1.4, a linear distribution isotherm was assumed. This is valid for all adsorbents at sufficiently low analyte concentrations. The linear range in the isotherm is apt to be shorter for carbon than for resin- and bonded-phase silica particles.

At 5% breakthrough, $(J/J_0) = 0.05$, $\tau$ is approximately 0.5 for $n = 5$ and approximately 0.7 for $n = 25$ (Fig. 1.4). This information can be used to estimate the breakthrough volume ($V_B$) for solid-phase extractions of given

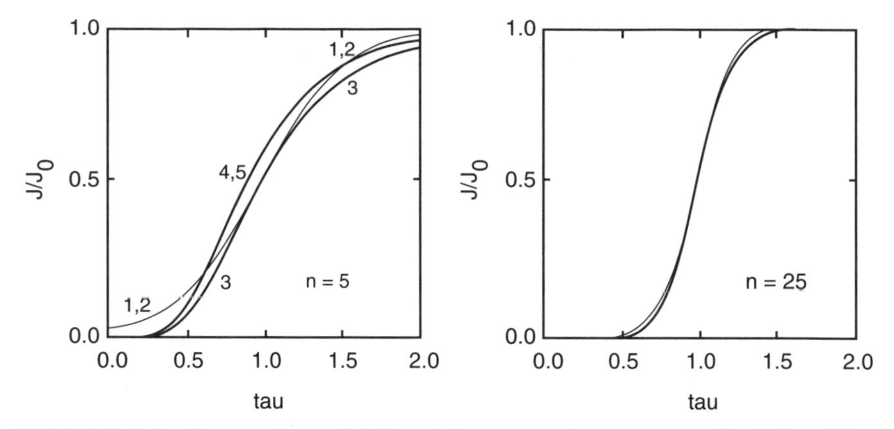


**FIGURE 1.4** Comparison of different front equations as normalized flux ($J/J_0$) versus relative retention time $\tau = t/t_R$ for 5 and 25 theoretical plates. (From Ref. 8 with permission.)

dimensions. For example, consider a column $L = 10$ mm, inner diameter (ID) = 5 mm and a void fraction of the packed column = 0.33. The void volume ($V_0$) of this packed column is $\pi$ (2.5/$^2$(10)(0.33) = 65 mm$^3$ = 0.065 ml. For an analyte with a capacity factor $k = 1000$, the retention volume $V_R$ = 0.065 (1001) = 65 ml. If the column has five theoretical plates, the breakthrough volume ($V_B$) at ($J/J_0$) = 0.05 is 0.5 × 65 ml = 32.5 ml. If the column has 25 theoretical plates, $V_B = 0.7 \times 65 = 45.5$ ml.

If the analyte retained on the column can be subsequently eluted by a small volume of an organic solvent, a considerable degree of concentration is possible. In this example it would be reasonable to assume that three void volumes of an organic solvent could elute the analyte completely: 3 $V_0$ = 3 × 0.065 = 0.2 ml. The maximum degree of concentration for $n$ – 25 is 45.5/0.2, or 227-fold.

In SPE of trace organic analytes from predominately aqueous samples, the equilibria do strongly favor partition into the solid phase in most cases. The capacity factor will, of course, vary with the structure of the analyte and the type of solid particles used, but $k$ = 1000–10,000 or even higher is often possible.

The number of theoretical plates in a SPE column is rarely measured as such, but in most cases it is probably small. Miller and Poole (9) found that the number of plates in a typical SPE cartridge with a bed height of 6 mm varied from $N = 25$ at a flow rate of 1 ml/min to $N = 9$ at a flow rate of 30 ml/min. Commercial cartridges at that time had low and variable packing densities, which sometimes led to poor performance with some channeling.

The number of plates in a given column length, and hence the efficiency of a SPE column, can be enhanced by careful packing and by use of adsorptive particles of small, uniform particle size.

## 1.3 ADVANTAGES OF SPE

Solid-phase extraction has several important advantages over liquid–liquid extraction:

1. *Faster; Easier Manipulation.* A sample can be quickly passed through a SPE column or cartridge by means of a pump or with gentle pressure or suction. After a quick rinse, the extracted substances can be washed from the column by a small volume of an organic solvent or another appropriate eluent. These steps can be automated readily. By contrast, simple solvent extraction requires a considerable amount of manipulation in adding the extractive liquid, shaking, waiting for the emulsion to break, and carefully separating the two liquid phases. Often a washing step and a backextraction are required. Although robotics can sometimes be applied, the steps involved in liquid–liquid extraction are inherently more complicated than those in SPE.

2. *Much Smaller Amounts of Liquid Organic Solvents Used.* The large quantities of organic solvents used in analytical separations have become an important environmental concern. Aqueous samples become contaminated with organic solvents and evaporative concentration of the extracts pollutes the air with organic vapors. Proper disposal of used organic solvents has become troublesome and expensive. The U.S. Environmental Protection Agency (EPA) is working to replace liquid–liquid extraction with solid-phase extraction in analytical procedures. It is difficult to completely remove all organic and metal impurities from organic liquids. When relatively large volumes of organic solvents are used, the sample extract will be contaminated with these impurities.

3. *Less Stringent Requirements for Separation.* SPE is a multistage separation method and as such requires only a reasonable *difference* in extractability to separate two solutes.

4. *Higher Concentration Factors.* The concentration factor is how many times more concentrated a substance is in the extract than it is in the original sample. In SPE, concentration factors of 1000 or more are possible. The concentration factor in liquid–liquid extraction depends in part on the volume ratio of the two liquids. Using vortex mixing it may be possible to

extract a 100-ml sample with perhaps 1 ml of an organic solvent. This could give a concentration factor of 100. However, the distribution ratio of the solute would need to be very high ($\sim 10^4$) to obtain complete extraction under these conditions.

## REFERENCES


1. W. Pipkin, *Am. Lab.* (Nov. 1990) 40D.
2. A. K. Burnham, G. V. Calder, J. S. Fritz, G. A. Junk, H. J. Svec, and R. Willis, *Anal. Chem.* **44** (1972) 139.
3. S. K. Poole, T. A. Dean, J. W. Oudsema, and C. F. Poole, *Anal. Chim. Acta* **236** (1990) 3–42.
4. M. Zief and R. Kiser, *Am. Lab.* (Jan. 1990) 70–82.
5. M. S. Mills and E. M. Thurman, *J. Chromatogr.* **629** (1993) 1–93.
6. N. Simpson, *Solid Phase Extraction: Principles, Strategies, and Applications*, Marcel Dekker, New York, 1997.
7. E. M. Thurman and M. S. Mills, *Solid-Phase Extraction, Principles and Practice*, Wiley, New York, 1998.
8. P. Lövkist and J. A. Jönsson, *Anal. Chem.* **59** (1987) 818.
9. K. G. Miller and C. F. Poole, *J. High Resol. Chromatogr.* **17** (1994) 125.

# CHAPTER 2

# SPE IN THE 1970s: EXTRACTION OF ORGANIC POLLUTANTS FROM WATER

## 2.1 INTRODUCTION

The early to mid-1970s was a period of rapid growth and development for solid-phase extraction. Actually, the technique was not given the name *solid-phase extraction* during that period. Solid particles such as activated carbon and porous polymeric resins were known as *sorbents* for trace organic constituents in aqueous samples. Columns filled with a solid sorbent were sometimes called *accumulation columns*.

Although the potable water supplies of cities and towns were commonly analyzed for inorganic constituents and biological organisms, very little was known in 1970 about the organic compounds naturally present in water or that were present because of some contamination. C. C. Johnson, on behalf of the American Public Health Association, stated in May 1, 1971 (1): We are considerably concerned about the amounts of chemical materials such as the wide spectrum of organics including pesticides and other chlorinated organics which may find their way into our water supplies and about the removal of these noxious agents from our water supplies. These similar concerns are held for those waters used in food processing and production.

In 1972 (2) Burnham et al. published the results of an investigation of the pollutants in the water supply of Ames, Iowa, which had an undesirable taste and odor when water from certain wells was used. The offensive material had been tentatively identified as phenols and cresols. This conclusion was based on a marginally positive 4-aminoantipyrene color test. A preconcentration of the offending chemicals was necessary to increase their concentration to a level at which their analytical detection and identification might be possible. The method developed was to pass 150 liters of water through a column (1.5 × 7.0 cm) filled with a porous polymeric resin (Rohm and Haas XAD-2). The flow rate through the column was 50 ml/min (4.0 bed volumes/min). The extracted organic pollutants were eluted from the column with 15 ml of ethyl ether. The ether effluent was carefully evaporated to a small volume.

Injection of a portion of the ether effluent into a gas chromatograph gave a number of well-separated peaks, corresponding to the organic pollutants that had been extracted from the water. However, identification of the peaks presented a real problem. Fortunately, the investigators had one of the first gas chromatography–mass spectroscopy (GC-MS) setups in existence and were able to identify the compounds listed in Table 2.1 (2).

The compounds identified are consistent with the source of contamination, which was believed to be residues from a coal gas plant operated in the city of Ames in the 1920s. The tar residues were buried in a pit that was connected hydrologically to an aquifer supplying the city water.

**TABLE 2.1 Neutral Compounds in a Contaminated Ames, Iowa Well**

| Name of Component | Concentration ppb | Standard Deviation |
|---|---|---|
| Acenaphthylene | 19.3 | 1.4 |
| 1-Methylnaphthalene | 11.0 | 0.6 |
| Methylindenes (two isomers) | 18.8 | 0.8 |
| Indene | 18.0 | 1.5 |
| Acenaphthene | 1.7 | 0.2 |
| 2-2-Benzothiophene | 0.37 | 0.11 |
| | | |
| Isopropylbenzene | | — |
| Ethylbenzene | | — |
| Naphthalene | | — |
| 2,3-Dihydroindene | 15[a] | |
| Alkyl-2,3-dihydroindene | | — |
| Alkyl benzenes | | — |
| Alkyl benzothiophenes | | — |
| Alkyl naphthalenes | | — |

## 2.2 COMPARISON OF ACTIVATED CARBON AND POROUS RESIN ADSORBENTS

Prior to 1972, the standard method for concentrating low concentrations of organic compounds from water samples was the carbon adsorption method (CAM). The severe limitations of this method have been documented (3).

One aspect of the (analytical) problem is that the carbon adsorption method recovers only a low percentage of the organic substances from water. It does not adsorb all the organics present; moreover a large portion of those adsorbed are not removed by the chloroform extraction. This indicates the need for a better method to collect and measure the organics in drinking water.

A second problem in the area of organic sampling is the recovery of the organics from water in unchanged form. The CAM can be used as a gross indication of the concentration of organics in water, but the residues recovered are not always in their original form, but may be altered by the use of solvents and heat. A need exists to recover the organics in an unchanged form if they are to be identified and used in determining their toxicity via animal experiments.

Later studies presented extensive data comparing the recoveries of organic compounds from water by carbon adsorption and resin adsorption methods (4,5). The results of one study are summarized in Table 2.2 (5). These results clearly suggest that organic polymeric resins are superior to activated carbon for isolating organic compounds from water.

Low recoveries can result from either breakthrough (incomplete extraction) or incomplete elution. It is possible that the latter may be the cause of some of the low recoveries in Table 2.2.

## 2.3 PURIFICATION OF RESINS

It is vital that the resins used as sorbents be as free as possible from impurities that might be leached from the solid particles during the elution step. The Rohm and Haas resins (XAD-2 and XAD-4) that were used during the 1970s often had to be ground up and sieved to obtain a particle size sufficiently small for efficient column extraction. The grinding of spherical resin particles exposed new resin surfaces to the leaching action of the organic solvent used in the elution step.

The recommended purification method was to purify the sized resin particles by sequential extraction with methanol, acetonitrile, and diethyl

**TABLE 2.2 Comparative Recovery Efficiency for Carbon and Resin Sorption Methods**

| Compound Type | Number of Compounds | | Carbon | XAD-2 Resin |
|---|---|---|---|---|
| Alcohols | 2 | | 48 | 100 |
| Aldehydes + ketones | 3 | | 4 | 74 |
| Alkanes | 5 | | 13 | 5 |
| Amines | 3 | | 24 | 14 |
| Aromatics | 7 | | 6 | 68 |
| Benzothiazoles | 2 | | 2 | 67 |
| Esters + ethers | 9 | | 52 | 74 |
| Halogenated compounds | 9 | | 32 | 57 |
| PCBs | 9 | | 12 | 78 |
| Phenols | 9 | | 10 | 46 |
| Total | 58 | Average percent recovery | 22 | 59 |

ether in a Soxhlet extractor for 8 h per solvent (6). Another effective method was to purge a heated resin-packed tube with nitrogen, with several injections of water to produce steam. The purified resins were stored in glass-stoppered bottles under methanol to maintain their high purity. In more recent years, polymeric resins for SPE are more pure and do not require the extensive treatment of the earlier materials. Still, blanks should always be run, especially for ultra trace work.

## 2.4 SAMPLING AND SAMPLE HANDLING

Figure 2.1 is a scale drawing of a sample extraction device. Grab samples were taken of surface waters using clean 4-liter amber solvent bottles. After settling overnight, the clear water was decanted into the reservoir shown in the figure. The stopcock (*E*) was adjusted to deliver a flow rate of 25–50 ml/min. When the water level reached the upper glass–wool plug, the sediment from the bottle was transferred to the reservoir using several rinses with organic-free water. In this way organic solutes adhering loosely to the sediment would be washed into the column containing the extractive resin.

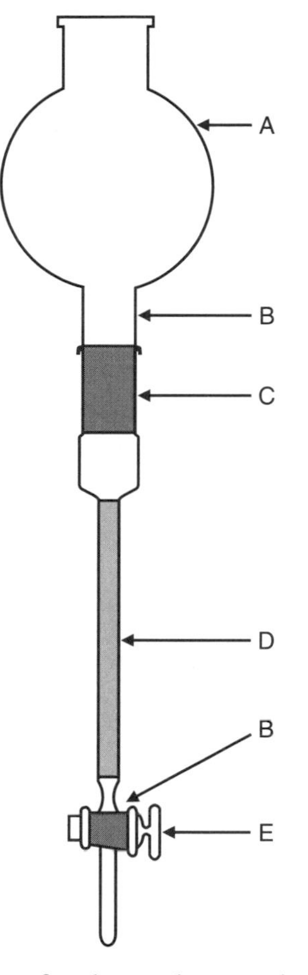


**FIGURE 2.1** Scale drawing of grab-sample extraction device: (*A*) 5-liter reservoir, scaled down; (*B*) glass–wool plugs; (*C*) 24/40 ground-glass joint with PTFE sleeve; (*D*) 8 × 140-mm glass tube packed with ~2 g 40–60-mesh resin; (*E*) PTFE stopcock.

Proper handling of water samples is also vital to the success of the method. Actual samples should be collected in a very clean glass bottle and analyzed without delay. During the passage of the sample through the resin column, the sample reservoir should be capped to prevent loss of some organics that tend to come to the water surface and be lost by evaporation. For example, 10 ppb (parts per billion) of cumene added to water samples showed only a 40% recovery with an uncapped reservoir but a 92% recovery with a capped reservoir. Water samples containing even low concentrations of saturated hydrocarbons present a special problem because some hydrocarbons cling tenaciously to the surfaces of the reservoir and column. It appears that this problem is alleviated by addition of either a detergent or a water-miscible organic solvent to the water sample.

The need for clean containers for water samples must be emphasized. In one experiment the recoveries of organic compounds in a standard sample were only about one half the usual recoveries. The difficulty was traced to a small grease spot inside the 4-liter sample bottle. On standing, some of the organic solutes were extracted from the water into the grease spot. Even a fingerprint can contain enough grease to take up some of the organic solutes.

The time at which a sample is taken and analyzed can affect the results. In a study on the determination of haloforms in drinking water, the amounts of $CHCl_3$, $CHBrCl_2$, and $CHBr_3$ were found to increase on storing the sample for several days. Continuing production of haloforms occurs while

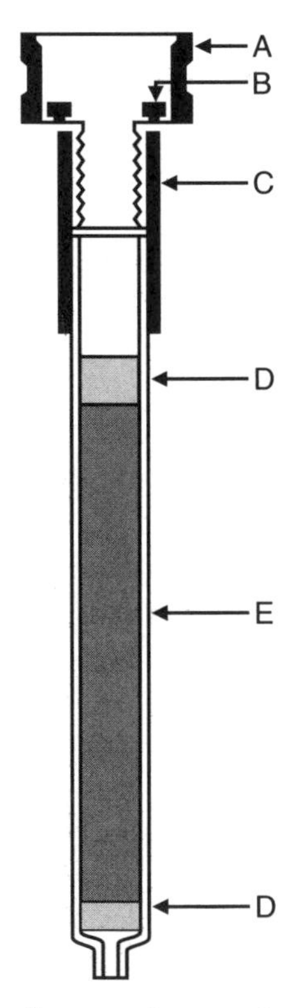


**FIGURE 2.2** Scale drawing of composite sample extraction device: (*A*) standard garden hose coupling; (*B*) PTFE washer; (*C*) 12.7-mm-ID PTFE tubing; (*D*) glass–wool plugs; (*E*) 12.7-mm-OD × 9-cm-long glass tube packed with ~2-g 40–60-mesh resin.

chlorinated water is in the distribution system. The increase in total haloform concentration with time suggests that the water samples still contain certain organic matter that reacts with residual chlorine or bromine to produce additional halocarbons (8).

The apparatus shown in Figure 2.2 was used for composite sampling over a ~24-h period (7). The standard garden hose coupling was attached to a suitable faucet and the water flow adjusted to deliver ~150 ml/min. After 24 h the column was removed from the coupling, and PTFE sleeves were used to connect a 25-ml reservoir and a PTFE stopcock to appropriate ends of the column. The column was then eluted with diethyl ether, and the eluate was treated as just described for the grab samples. Either sampling procedure may be employed dependent on whether one is interested in instantaneous (grab) or average (composite) concentrations over an extended time period.

The composite sampling device was initially connected to a faucet by a short length of garden hose. However, the use of even short lengths of polyvinylchloride tubing was found to introduce significant amounts of organic impurities. Some of these contaminants appeared to be plasticizers leached from the PVC tubing. This led to a study of the organic contamination from the different types of tubing used to contain or transfer various fluids (3).

## 2.5 PREPARATION OF STANDARDS

Water samples containing known amounts of various organic analytes are needed to check out a concentration/analysis method and to determine the percentage recovery for the analytes. Since the solubility of many organic compounds is quite low, a technique must be used that will quickly dissolve the organic compound and give a homogeneous solution to sample. The best way to do this is first to prepare a standard solution of each organic test compound. This can be a solution of perhaps 150 ppm of the test compound in a solvent such as methanol, acetone, or ethyl acetate. Huge losses can occur when hexane is the solvent, probably because of a floating layer and subsequent vaporization. A small but measured volume of this standard solution can be added to pure water just before adding the water sample to the SPE column. For example, 100 μl of standard solution (150 ppm) added to 15 ml of water produces a sample containing 1 ppm of the test compound. It is best to "squirt" the standard solution into the water as quickly as possible in order to provide some immediate mixing. With this technique any transient precipitate will consist of very small particles that will quickly dissolve as equilibration becomes more complete.

The organic solvent for the 150-ppm standard solution should be selected with care. Any solvent that might undergo a chemical reaction with the analyte should be avoided. Fresh standard solutions should be prepared from time to time.

## 2.6 GENERAL PROCEDURE FOR SPE

The procedure given below (6) is typical of those used for solid-phase extractions in the 1970s. Today, smaller, more efficient resins, small columns, and different eluting solvents are used. However, the research of the 1970s provides a sound foundation for SPE as it is performed today. A comprehensive paper published in 1974 (6) has been identified as one of the 20 most frequently cited papers published in the *Journal of Chromatography* (9).

### 2.6.1 Column Preparation

The apparatus used was similar to that in Figure 2.1 except for a 2-liter reservoir instead of a 5-liter. With the upper 2-liter reservoir detached, insert a clean silanized glass–wool plug near the stopcock of the glass column. Add the purified XAD resin as a methanol slurry until a resin bed approximately 6 cm high is obtained (1.5–2.0 g of dry resin), then insert a second silanized glass–wool plug above the resin. Drain the methanol through the stopcock until the level just reaches the top of the resin bed, then wash the resin with three 20-ml portions of pure water. For each portion stop the flow when the liquid level reaches the top of the resin bed.

### 2.6.2 Sample Preparation

Attach the 2-liter reservoir to the column (Fig. 2.1) and add 1000 ml of purified distilled or tap water. Add the organic compound(s) to be tested to the water by injecting a calibrated volume of a standard solution of the organic compound(s) in methanol. The volume injected and the concentration of the standard is adjusted to achieve the desired amount of compound(s) in 1 liter of water.

### 2.6.3 Column Extraction

Cap the reservoir with a one-hole stopper connected to a nitrogen source and allow the water to pass through the XAD resin column by gravity flow at a rate of 30–50 ml/min. If the flow rate is slower than this, apply a pressure of about 1 psi (lb/in.$^2$) to the reservoir using organic-free nitrogen. When most of the sample has passed through the column and the liquid level is at the top of the resin, wash the reservoir walls carefully with a 20-ml portion of pure water and drain through the column until the level reaches the top of the resin bed. Repeat this wash twice, letting the water drain completely only after the last wash.

### 2.6.4 Elution and Regeneration

Wash the reservoir walls with two 10-ml portions of diethyl ether and allow each wash to drain into the XAD resin but not through the column. Remove the reservoir, cap the column with a 24/40 glass stopper, and allow the diethyl ether to equilibrate with the resin for 10 min. Then remove the cap, open the stopcock, and allow the ether to flow through the column into a 30-ml test tube. Add an additional 5 ml of diethyl ether to the column and immediately allow it to flow through the resin into the receiver. The elution is complete when the gravity flow ceases even though the last traces of ether have not been removed from the resin.

Regenerate the XAD resin column immediately after the ether elution. Add methanol, shake to remove any air bubbles, then pass a total of 30 ml of methanol through the column. Close the stopcock, add an additional 15 ml of methanol, and cap the column with a 24/40 glass stopper. It is now ready for subsequent analyses without any further treatment beyond wetting the resin with pure water as outlined above in step 1.

### 2.6.5 Drying

Remove the residual water (0.5–2 ml) from the diethyl ether eluate by immersing the test tube receiver in liquid nitrogen for two 10-s intervals. Immediately decant the ether into the concentration vessel. Wash the ice in the test tube with 1 ml of diethyl ether, dip briefly in the liquid nitrogen to freeze any ice that may have melted and add this ether to the concentration vessel.

Alternatively, anhydrous sodium sulfate and petroleum ether (b.p. 30–60) may be used to dry the ether eluate. Collect the ether eluate from the column in a 60-ml separatory funnel and reject the water layer. Add 15 ml of petroleum ether (30–60°) and 2 g of anhydrous sodium sulfate to the ether. Shake vigorously until a clear solution is obtained and transfer the ether to the concentration vessel.

### 2.6.6 Concentration of Eluate

Add a small boiling chip to the concentration vessel and attach a three-cavity Snyder column. Add about 2 ml of diethyl ether to the top of the Snyder column and tap gently to distribute the ether into the three cavities. Apply heat from a hot plate or steam bath so that the boiling action is vigorous enough to agitate the balls of the Snyder column continuously. A solvent evaporation rate of 0.5–2.0 ml/min should be attained. When the volume of solution in the calibrated appendage of the concentration vessel is about 0.5 ml, remove the apparatus from the heat and immediately spray acetone over the outside walls of the concentration vessel. The condensation of the ether vapor causes an automatic sequential washing of the inside walls of the vessel with the ether held in the three cavities of the Snyder column. The volume of liquid in the calibrated section of the concentration vessel should now be 1.0 ml. Remove the Snyder column and add ether if necessary so that the solution volume is exactly 1.00 ml. Cap the vessel with a 14/20 stopper and swirl to mix the solution. Proceed with the GC analysis of the concentrate as soon as possible.

The shape of the vessel used to concentrate the eluate is critical. Vessels of several shapes were tested (Fig. 2.3). The calibrated scale at the bottom is for measurement of the volume of concentrated eluate. A vessel of shape

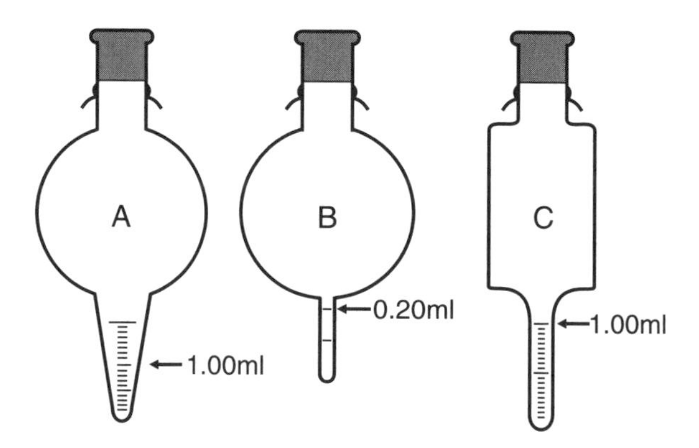


**FIGURE 2.3** Drawings of the concentration vessels: A is recommended, B is unsatisfactory, and C is questionable. (From Ref. 6 with permission.)

*A* was found to give the best results. Vessels of shape *B* or *C* were found to cause losses of 10% to 60% during the evaporation step.

### 2.6.7 Separation and Quantification

Inject a suitable aliquot of the 1.00-ml concentrate into the gas chromatograph with a syringe. When the GC separation of the organic compound(s) in the concentrate is completed, immediately repeat the GC separation using a 2.0-μl aliquot of a 1.00-ml standard ether solution containing the same organic compound(s) at a concentration identical to that expected in the 1.00-ml concentrate assuming complete recovery of solute(s) from the water sample. The GC conditions are held rigidly constant for both sample and standard during these tests, and the percentage recovery of the organic solutes is calculated directly from a comparison of the chromatogram peak heights.

### 2.6.8 Identification

Further concentrate the 1.00-ml ether solution to 0.1 ml after completion of step 7. Subject a 2.0-μl aliquot of this 0.1-ml solution to GC-MS analysis to confirm positively that no chemical transformation has occurred during any of the previous steps.

**TABLE 2.3 Average Recoveries of Test Compounds by Chemical Groups Using the "Porous Polymer" Extraction Method**

| Group | Number of Compounds | Recovery, % |
|---|---|---|
| Acids | 5 | 101 |
| Alcohols | 8 | 95 |
| Aldehydes | 7 | 95 |
| Alkylbenzenes | 3 | 89 |
| Esters | 15 | 93 |
| Ethers | 5 | 90 |
| Halogen compounds | 10 | 87 |
| Nitrogen compounds | 10 | 89 |
| Pesticides | 6 | 83 |
| Phenols | 7 | 82 |
| Polynuclear aromatics | 8 | 88 |

*Source:* Data are summarized from Reference 6.

## 2.7 RECOVERY DATA

The recovery efficiency of the entire analytical scheme known as the *porous polymer method* was studied extensively and the results reported in a comprehensive paper published in 1974 (6). Organic test compounds were added to distilled water, and in some cases to tap water, in concentrations ranging from 10 to 100 ppb. The test compounds included acids, alcohols, aldehydes, alkylbenzenes, aromatic halides, esters, ethers, ketones, phenols

**TABLE 2.4 Recoveries of Some Individual Compounds by Porous Polymer Extraction.**

| Test Compound | Recovery, % | Test Compound | Recovery, % |
|---|---|---|---|
| Acids | | Ethers | |
| Benzoic | 107 | Hexyl | 75 |
| Oleic | 100 | Anisole | 87 |
| Alcohols | | Halogen compounds | |
| Hexyl | 93 | Benzyl chloride | 88 |
| Decyl | 91 | Chlorobenzene | 95 |
| Benzyl | 91 | 1,2-Dichlorotoluene | 96 |
| Cinnamyl | 85 | 2,4-Dichlorotoluene | 71 |
| Aldehydes | | Nitrogen compounds | |
| Benzaldehyde | 101 | Nitrobenzene | 91 |
| Salicylaldehyde | 100 | Indole | 89 |
| | | Quinoline | 84 |
| Alkyl benzenes | | Pesticides | |
| Ethylbenzene | 81 | Atrazine | 83 |
| Cumene | 93 | Lindan | 95 |
| Esters | | Phenols | |
| Dimethylphthalate | 91 | Phenol | 40 |
| Dibutylphthalate | 92 | *o*-Cresol | 73 |
| Methylbenzoate | 101 | 2-Chlorophenol | 96 |
| Methyloctanoate | 98 | 2,4,6-Trichlorophenol | 99 |
| Methylpalmitate | 70 | | |
| | | Polynuclear aromatics | |
| | | Naphthalene | 98 |
| | | Anthracene | 83 |
| | | Acenaphthene | 92 |

*Source:* This is an abbreviated list from Reference 6.

and chlorinated phenols, pesticides and herbicides, polynuclear aromatics, and various compounds containing halogens, nitrogen, or sulfur. Aqueous samples containing acidic solutes were acidified with 5 ml/liter of concentrated hydrochloric acid to ensure that the compounds were in their molecular form.

A summary of the average recoveries obtained for 84 individual compounds are summarized according to compound class in Table 2.3. The average recovery for the 84 compounds studied was 90%.

The recoveries of an abbreviated list of individual compounds are shown in Table 2.4. These results show the effect that chemical structure can have on the recoveries. For example, phenol itself tends to have a significantly lower recovery than phenols that contain alkyl or halogen groups.

## REFERENCES

1. A. K. Burnham, G. V. Calder, J. S. Fritz, G. A. Junk, H. J. Svec, and R. Vick, *J. Am. Waterworks Assoc.* **65** (1973) 722.
2. A. K. Burnham, G. V. Calder, J. S. Fritz, G. A. Junk, H. J. Svec, and R. Willis, *Anal. Chem.* **44** (1972) 139.
3. G. A. Junk, H. J. Svec, R. D. Vick, and M. J. Avery, *Env. Sci. Technol.* **8** (1974) 1100.
4. C. D. Chriswell, R. L. Ericson, G. A. Junk, K. W. Lee, J. S. Fritz, and H. J. Svec, *J. Am. Waterworks Assoc.* **69** (1977) 669.
5. G. A. Junk, C. D. Chriswell, R. C. Chang, L. D. Kissinger, J. J. Richard, J. S Fritz, and H. J. Svec, *Z. Anal. Chem.* **282** (1976) 331.
6. G. A. Junk, J. J. Richard, M. D. Grieser, D. Witiak, J. L. Witiak, M. D. Arguello, R. D. Vick, H. J. Svec, J. S. Fritz, and G. V. Calder, *J. Chromatogr.* **99** (1974) 745.
7. G. A. Junk, J. J. Richard, H. J. Svec, and J. S. Fritz, *J. Am. Waterworks Assoc.* **68** (1976) 216.
8. L. D. Kissinger and J. S. Fritz, *J. Am. Waterworks Assoc.* **68** (1976) 435.
9. J. S. Fritz and G. A. Junk, *J. Chromatogr.* **625** (1992) 87.

CHAPTER 3

# SOLID PARTICLES FOR SOLID-PHASE EXTRACTION OF ORGANIC COMPOUNDS FROM WATER

## 3.1 INTRODUCTION

A large number of solid particles have been used for SPE of organic compounds from predominately aqueous samples. Bonded-phase silica particles, especially ODS silica (octadecylsilane silica), are currently the most popular type. However, a wide variety of porous polymeric resins have also been used. Activated carbon and more recently graphitized carbon materials constitute another class of solid particles for SPE.

Before discussing the actual particles used in solid-phase extraction of organic compounds, let us review the various types of SPE available. Classification of the types of SPE available for organic analytes is the same as for HPLC, although the end goal of SPE is different from HPLC (see Chapter 1).

1. *Reversed-Phase SPE.* Here the goal is to isolate relatively nonpolar analytes from a polar sample such as water. This type of application requires the use of relatively hydrophobic adsorbent particles, such as silica with bonded octadecylsillane groups or an organic polymer

with benzene rings. The extracted substances are eluted by a small volume of an organic solvent.

2. *Normal-Phase SPE.* This technique is used to isolate polar compounds from a nonpolar sample matrix. A vegetable oil dissolved in hexane is an example of such a sample matrix. Polar solid particles are then used to extract polar analytes from the sample. The extracted analytes are finally eluted by a polar solvent.
3. *Ion-Exchange SPE.* Particles containing cation-exchange groups or anion-exchange groups are used to extract ionic analytes or analytes that can be converted to ionic form by adjusting the sample pH. After the SPE step, the extracted substances are eluted with an organic solvent after converting them back to the molecular form. Alternatively, the extracted ions may be removed from the ion-exchange sites by elution with a solvent containing a relatively high concentration of a displacing ion. SPE with ion-exchange particles is discussed in Chapter 5.

In this chapter the desirable properties of SPE materials will be listed, followed by a discussion of the many types of particles that are used in solid-phase extraction.

## 3.2 DESIRABLE PROPERTIES OF SPE PARTICLES

### 3.2.1 Porous, Large Surface Area

In SPE the uptake of a sample solute depends on an equilibrium between the sample solution and the solid SPE particle. This equilibrium is shifted more strongly toward the solid as the surface area becomes larger. A surface area >100 $m^2/g$ is generally necessary for effective SPE. The particles most widely used today have a surface area between about 200 and 800 $m^2/g$.

The pore size of a particle and its surface area are generally linked in an inverse relationship. The surface area decreases as the average pore size increases. This relationship, together with the ability of a series of polymeric adsorbents to extract a copper(II)–dithiocarbamate complex from water is illustrated by the data in Table 3.1 (1). In both the polystyrene resins and the polyacrylate resins the extractability of the copper organic complex increases with surface area.

**TABLE 3.1 Physical Characteristics and Breakthrough Capacities of XAD Resins for CopperII–Di(hydroxyethyl)dithiocarbamate**

| Adsorbent | Type | Surface Area, $m^2/g$ | Pore Size Å | Capacity, mm Cu/g |
|---|---|---|---|---|
| XAD-1 | Polystyrene/DVB | 100 | 100 | 0.027 |
| XAD-2 | Polystyrene/DVB | 300 | 90 | 0.132 |
| XAD-4 | Polystyrene/DVB | 725 | 40 | 0.474 |
| XAD-7 | Polyacrylate | 450 | 90 | 0.239 |
| XAD-8 | Polyacrylate | 160 | 225 | 0.132 |

### 3.2.2 Reversible Adsorption

Successful SPE has two major requirements: (1) a high, reproducible percentage of the analytical solutes must be taken up by the solid extractant; and (2) the solutes must then be easily and completely eluted from the solid particles. Carbon was the first medium to be used for the extraction of organic compounds from water (2). Carbon has the advantage of very high surface area and consequently high uptake of organic solutes. However, the heterogeneous nature of the activated carbons used for SPE has caused problems such as irreversible sorption, widely differing affinities of different classes of compounds, and catalytic activity of the surface, which can lead to unwanted chemical reactions.

Other widely used SPE materials such as bonded-phase silicas and porous polymers generally pose no problems with regard to reversible adsorption. A number of organic solvents are available that can effectively elute adsorbed solutes from the solid particles.

### 3.2.3 Pure, Low Leachable Impurities

The solid particles need to be as free as possible of impurities that might be leached from the solid during elution of the retained sample constituents. If purification of the particles is needed, it often can be accomplished by washing with an organic solvent or a succession of solvents.

The XAD resins used in the 1970s and 1980s posed a rather severe purification problem. Ethylbenzene, benzoic acid, and other impurities were trapped within the resin during the polymerization step. At that time, crushing and sieving was necessary to reduce the materials to a particle size suitable for solid-phase extraction. This process opened up additional surfaces from which the resin impurities continued to be leached. Fairly extensive Soxhlet extraction with several organic solvents was needed to reduce the leachable impurities to an acceptable level. Another purification method was to heat a tube containing resin in the oven of a gas chromatograph and to steam-clean by injecting several small portions of water.

Impurities can come from sources other than the particles used in SPE. Junk, Avery, and Richard (3,4) found that various compounds can be leached from the polypropylene housing often used in commercial SPE cartridges and also from the frit that holds the particles in place. Table 3.2 lists compounds leached from the cartridge housing, frit, and the C18-bonded porous silica. The impurities leached from commercial cartridges varies

**TABLE 3.2 Compounds Identified in Extracts of Components of Solid-Phase Extraction Cartridges**

| | Column Component | | |
|---|---|---|---|
| Compound | Polypropylene Frit | Polyethylene Frit | C18-Bonded Porous Silica |
| C8 alkene | —[a] | — | x |
| C9 alkene | — | — | x |
| C10–C16 alkenes | x | — | x |
| C17–C28 alkenes | x | x | — |
| C8–C26 alkanes | — | — | x |
| Naphthalene | | — | x |
| Biphenyl | x | — | — |
| Acenaphthene | x | — | — |
| 2,6-Di-*tert*-butyl-*p*-cresol | x | — | — |
| 2,6-Di-*tert*-butyl-*p*-quinone | x | — | — |
| Phenol | x | — | — |
| Methyloctadecanoate | x | — | — |
| Dimethyloctadecylsilanol | — | — | x |
| Decamethylpentasiloxane | — | — | x |
| Dodecamethylhexasiloxane | — | — | x |
| Tetradecamethylheptasiloxane | — | — | x |
| Diethylphthalate | — | — | x |
| Dibutylphthalate | x | x | x |
| Bis(2-ethylhexyl)phthalate | x | x | x |
| Bis(2-ethylhexyl)adipate | x | — | — |

[a]The dash (—) means not positively identified in extract.

*Source:* From Reference 3 with permission.

with the eluting solvent and often with the lot number of the cartridge (Table 3.3).

### 3.2.4 Good Chemical Stability

An acidic or basic sample solution or eluting solution is needed in some applications. Silica-base adsorbents are not very stable above pH 8 or in highly acidic solutions. However, most polymeric resins are quite stable in both basic and acidic media.

Instead of eluting adsorbed sample analytes with an organic solvent, it is often convenient to use thermal desorption. The SPE minicolumn can be placed in a convenient heater that is attached to a gas chromatograph and the analytes thermally desorbed directly into the chromatograph. This mode of operation requires the use of solid particles with good thermal stability.

Solid particles need to be stable in the organic solvents used in the elution step as well as in the aqueous sample. Polymeric resins should be suffi-

**TABLE 3.3 Compounds Eluted from Commercial C18 Cartridges**

| | Eluting Solvent | |
|---|---|---|
| Compounds | Ethyl Acetate | Benzene |
| C10 alkane | x | —[a] |
| C19 alkane | x | — |
| C20 alkane | x | — |
| C22–C18 alkanes | x | x |
| C15–C17 alkenes | — | x |
| C18–C26 alkenes | x | — |
| Dimethyloctadecylsilanol | x | — |
| Methylhexadecanoate | x | x |
| Hexadecanoic acid | — | x |
| Methyloctadecanoate | — | x |
| 2,6-Di-*tert*-butyl-*p*-cresol | — | x |
| Naphthalene | — | x |
| Diethylphthalate | — | x |
| Dibutylphthalate | — | x |
| Bis(2-ethylhexyl)phthalate | — | x |
| Bis(2-ethylhexyl)adipate | — | x |

[a]The dash means not positively identified in eluant.

*Source:* From Reference 3 with permission.

ciently crosslinked to ensure that they do not dissolve or soften in organic solvents. The resins must not undergo large volume changes due to swelling and shrinking in contact with different liquids.

### 3.2.5 Good Surface Contact with Sample Solution

Chemically bonded silica, usually with a C18 or C8 organic group is the most used material for SPE. However, the use of crosslinked polystyrene and other porous polymeric resins is increasing. Polymeric resins are more rugged and pH-stable, and they have a greater surface area than do most silica-based materials. Research has consistently shown that the percentage recovery for many types of analytes is significantly higher with polymeric resins than with silica particles (5).

Chemically bonded silica and porous polystyrene resins have several shortcomings for use in SPE (6). While silica itself is hydrophilic, the hydrocarbon chains make the surface hydrophobic. The consequence is poor surface contact with predominantly aqueous solutions. Porous polystyrene resins also have a hydrophobic surface (7). Pretreatment of the SPE materials with an activating solvent (such as methanol, acetone, or acetonitrile) must be used to obtain better surface contact with the aqueous solution being extracted. However, the activating solvent can be gradually leached out of the resin, thereby causing the extraction to become ineffective. This is particularly true if the SPE column inadvertently becomes dry, causing air to be sucked into the column (8).

A better approach is to make the surface of a solid-phase extractant permanently hydrophilic through a chemical reaction. Sun and Fritz (9,10) introduced an acetyl, hydroxymethyl, or cyanomethyl group into crosslinked polystyrene resins at a capacity of approximately 1 mmol/g. Whereas dry, untreated resins float and clump together on the surface of water, the modified resins have a more hydrophilic surface and are easily wetted by water. The derivatized resins can be used in SPE without any pretreatment with an "activating" solvent such as methanol.

Mild sulfonation of resins also increases their hydrophilicity and makes them wettable with water alone (11). The best performance for SPE was found to be with resins containing around 0.6 meq/g of sulfonate groups. The degree of sulfonation should be sufficient to render the resin surface hydrophilic, but not enough to compete with the more hydrophobic resin matrix, which is responsible for extraction of the organic analytes from water.

### 3.2.6 High-Percentage Recoveries

Quantitative analyses involving SPE depends on recovery of a known, fixed percentage of each analyte during the extraction and elution steps. Ideally, quantification will be based on essentially complete (~100%) recovery. Using known standards, quantification based on a lower percentage recovery is feasible, but precision and accuracy are better when the overall percentage recovery is as high as possible.

## 3.3 BONDED-PHASE SILICA PARTICLES

### 3.3.1 Types Available

At this writing bonded-phase (BP) silica materials are the dominant particles used in SPE. One reason for this is their widespread availability. When the advantages of SPE as a sample pretreatment technique became apparent and the use of SPE began to increase, commercial suppliers made products available to meet the new demand. Since the technology of bonded-phase silica packings for HPLC columns was already well developed, it was a relatively easy task to offer similar materials in a form suitable for SPE. As the use of BP silica materials for SPE increased, several supply houses offered bibliographies and other literature to users. It then became possible for chemists to use this information as a guide for specific analytical problems they might encounter.

Several supply houses offer a full line of silica particles for SPE. The major types are listed in Table 3.4. The materials in this table are listed in their approximate order (top to bottom) of increasing polarity. The desired major functional group is introduced into porous silica particles by reaction of silanol groups on the silica with chloro- or methoxyorganosilane. For example, the C8-bonded phase material has the following structure:

$$\text{Silica—OH} + \text{Cl–}\underset{\substack{|\\CH_3}}{\overset{\substack{CH_3\\|}}{Si}}\text{—}CH_2CH_2CH_2CH_2CH_2CH_2CH_2CH_3 \longrightarrow$$

$$\text{Silica–O—}\underset{\substack{|\\CH_3}}{\overset{\substack{CH_3\\|}}{Si}}\text{—}CH_2CH_2CH_2CH_2CH_2CH_2CH_2CH_3 \quad (3.1)$$

**TABLE 3.4 Bonded-Phase Silica-Modified Materials Used in Solid-Phase Extraction**

| Phase | Polarity of Phase | Designation |
|---|---|---|
| Octadecyl, endcapped | Strongly apolar | C18ec |
| Octadecyl | Strongly apolar | C18 |
| Octyl | Apolar | C8 |
| Ethyl | Slightly polar | C2 |
| Cyclohexyl | Slightly polar | CH |
| Phenyl | Slightly polar | PH |
| Cyanopropyl | Polar | CN |
| Diol | Polar | 2OH |
| Silica gel | Polar | SiOH |
| Carboxymethyl | Weak cation exchanger | CBA |
| Aminopropyl | Weak anion exchanger | $NH_2$ |
| Propylbenzene sulfonic acid | Strong cation exchanger | SCX |
| Trimethylaminopropyl | Strong anion exchanger | SAX |

The alkali chain in the octadecylsilane (ODS) material is even longer (18 carbon atoms in the alkyl chain). These particles may be visualized as porous solids with a high surface area and the alkylsilane sticking out from the various surfaces like tall trees. Actually the long alkyl chains have a kind of zigzag structure. It will be seen that longer alkyl chains make the silica particles more hydrophobic and also provide a stronger overlap with the hydrophobic parts of extracted analyte molecules. For this reason, extractibility generally increases with chain length.

The bulk of the chloro alkyl silanes used in the derivatization reaction prevents their reaction with all of the SiOH groups on the silica. These free silanol groups can attract polar analytes that might not otherwise be extracted by hydrogen bonding. Alcohols and amines are examples:

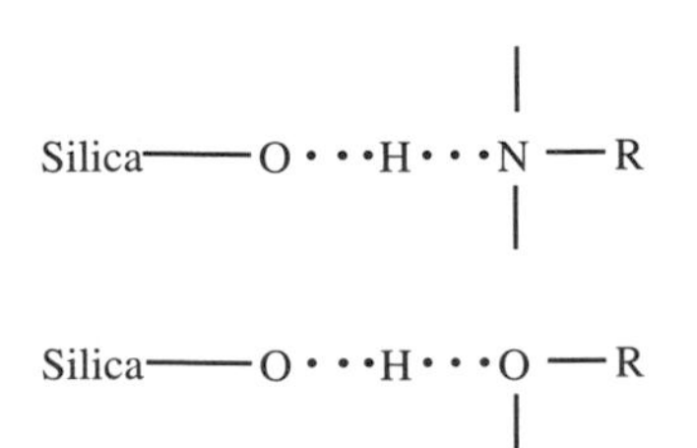

To avoid this situation, a process known as *endcapping* is often used. The remaining silanol groups are endcapped by reaction with a smaller, more reactive organosilane such as chlorotrimethylsilane.

$$\text{Silica–OH} + \text{Cl Si(CH}_3)_3 \rightarrow \text{Silica–O–Si(CH}_3)_3 \qquad (3.2)$$

Octadecylsilane silica, often abbreviated as ODS or C18, is by far the most widely used bonded-phase silica for reversed-phase SPE. Octylsilane (C8), ethylsilane (C2), and occasionally cyclohexylsilane (CH) also find some application, especially for SPE of more bulky analytes. Phenyl-substituted silica may be used for reversed-phase extraction of somewhat more polar analytes. Cyanopropyl silica and silica with a diol bonded phase are generally too polar to be used in the reversed-phase mode. Instead, these materials are often used for SPE in the "normal-phase" extraction mode. Silica gel and several other inorganic particles are also used for normal-phase SPE (see Section 3.8). The last four BP silicas listed in Table 3.3 are used in ion-exchange SPE (Chapter 5).

Bonded-phase silicas are chemically stable in the presence of most organic solvents. The average pore size of approximately 60 Å makes it possible to extract organic compounds of up to ~15,000 molecular weight. Sample solutions much above pH 8 cause some hydrolysis of BP silica particles.

### 3.3.2 Cartridges

SPE with BP silica particles may be carried out in pre-prepared cartridges, in prepacked tubes of varying dimensions, or in various devices in which the silica is incorporated in a disk (Chapter 6). The basic design of cartridges has changed little since their introduction in the late 1970s. A typical cartridge consists of a polyethylene body with a female luer tip at the top for attachment to a positive-pressure source and a male luer tip at the bottom. The packing material is held in place by a 20-μm polyethylene frit at each end. The particle size of the packing material varies, but typically averages around 40 or 50 μm in diameter. The dimensions of the sorbent bed are small enough to permit easy flow of the sample through the cartridge, either by gravity or by suction-aided flow. The small dimensions also minimize the volume of organic solvent needed to condition the column and elute the sorbed analytes.

Properties of two batches of a popular SPE cartridge are summarized in Table 3.5. The C18 sorbent has small pores and a high surface area. It has a heavy loading of surface-bonded C18 groups. The particle size range is

**TABLE 3.5 Sorbent Properties of C18 Silica Extraction Cartridges as Determined by HPLC**

| | J. T. Baker Octadecylsiloxane Sorbent | |
|---|---|---|
| Parameter | Column 1 | Column 2 |
| Dimensions | 25 cm<br>2.1 mm ID | 25 cm<br>4.6 mm ID |
| Lot number | F37502 | G10508 |
| Weight of packing (g) | 0.570 | 3.3793 |
| Column volume (ml) | 0.785 | 4.155 |
| Packing density (g/ml) | 0.726 | 0.813 |
| Total porosity | 0.49–0.51 | 0.47 |
| Interparticle porosity | 0.45–0.42 | 0.41 |
| Intraparticle porosity | 0.06–0.07 | 0.06 |
| Apparent average particle diameter (μm) | 55.1 | 56.5 |

*Source:* Data from Reference 12.

much larger than that used in HPLC, and the distribution is skewed toward smaller particles.

Miller and Poole used HPLC to study the properties of typical octadecylsiloxane silica cartridges used for SPE (12). The characteristics of the HPLC columns used in their study are summarized in Table 3.5.

The total porosity of the column includes the liquid between the silica particles (interparticle porosity) plus the liquid within the pores of the particles. The latter is called the *intraparticle porosity*. The total porosity ($E_t$) is measured by injection of a nonretained chemical marker that can freely enter the pores within the particles.

$$E_t = \frac{F_v t_m}{\pi r^2 L} \tag{3.3}$$

where $F_v$ is the volume flow rate, $t_m$ is the time for the marker to elute, $r$ is the column radius, and $L$ is the length of the resin bed. The intercolumn porosity is calculated using dilute sodium nitrate as the marker because it is excluded from the intraparticle pores by ion exclusion. The intraparticle porosity is the difference between the total porosity and the interparticle porosity. It will be seen from Table 3.5 that the intraparticle porosity of the C18 silica is quite small. This may be the result of plugging many of the small pores due to the heavy loading of bonded octadecylsiloxane groups.

These results indicate that the material does not have an accessible internal pore structure.

The average packing densities of several SPE cartridges were found to be 0.667 g/ml ($n = 5$, $s = 0.014$), with a range of 0.653–0.683 g/ml. These packing densities are considerably smaller than those found for C18 HPLC columns in Table 3.5. This fact, together with the rather large spread in packing densities, suggests that the cartridge beds are not very stable. This heterogeneous packing structure of the cartridges leads to channeling and reduces their effectiveness for SPE.

### 3.3.3 SPE Tubes

Prepacked SPE tubes are very efficient and easy to use. They seem destined to replace the older cartridges. A typical SPE tube consists of a syringe barrel packed with 40–50-µm sorbent material. There is a male luer tip at the bottom and the packing is held in place by polypropylene frits. SPE tubes are fairly low in cost and are disposable. A wide variety of SPE tubes is available from chromatographic supply houses. For example, tubes packed with C18 silica are available in the following sizes: 1, 3, 6, 8, 12, 20, and 60 ml. The weight of sorbent in these tubes is 0.1, 0.5, 0.5, 1.0, 2.0, 5.0, and 10 g, respectively. Tubes and other devices for SPE are described more fully in Chapter 4. The availability of this equipment means that the scale of SPE (sample size, volume of eluting solvent) can be varied over a broad range.

Although they are pure, are convenient to use, and generally give acceptable results, bonded-phase silica materials do have several drawbacks. The need to pretreat the tube or cartridge with a solvent such as methanol was pointed out in Section 3.1.5. The recommended pH range for use of BP silica materials is generally specified as pH 2–8. Especially at highly alkaline pH values, the amount of organic coating is reduced by hydrolysis. This causes the particles to lose part of their extractive power and may result in unexpectedly low recoveries of sample compounds.

The overall recoveries (extraction + desorption) obtained with BP silica particles varies considerably. Recoveries of some analytes are close to 100%, while others have much lower recoveries. Published procedures often involve the use of internal standards for quantification and fail to give the actual recoveries of the various analytes. The advantages of high-percentage recoveries were pointed out in Section 3.2.6. The main point here is that BP silicas often give significantly lower overall recoveries than do other SPE materials, particularly polymeric sorbents.

## 3.4 ORGANIC POLYMERIC ADSORBENTS

Porous organic polymers have a number of advantages over activated carbons and bonded-phase silicas for solid-phase extraction. Unlike bonded-phase silicas, polymeric organic particles can be used at virtually any pH, and they contain no troublesome silanol groups. The surface area is generally higher than that of silica particles, and uptake of organic analytes thus tends to be more complete. In most cases adsorbed analytes are easily and completely eluted from polymeric adsorbents by a small volume of an organic solvent.

The use of polymeric adsorbents for extraction has been reviewed (13). The Porapak (Waters) and Chromasorb (Johns-Manville) adsorbents listed in this review have been available for some years, but they are infrequently mentioned in the more recent literature on SPE. Table 3.6 lists polymeric adsorbents that are finding extensive use in current practice. The XAD materials, developed by Rohm and Haas and now marketed by Supelco and other supply houses, have had a major impact on SPE. Their properties are given in Table 3.1, and some of their earlier uses are described in Chapter 2. Although these are excellent sorbents, in many cases they continue to be available only in a 20–60 mesh particle size range (approximately 400–250 μm). This is much too large for efficient solid-phase extraction and means that additional grinding and sizing will be required. Even when the original particles have been purified, grinding opens up new surfaces from which organic impurities can be leached over a long period of time. However, some newer versions of these resins (CG sorbents) are listed in Table 3.6 that have been purified and are available in a more suitable particle size range.

Several companies produce porous crosslinked polystyrene–divinylbenzene (PS-DVB) resins of smaller particle size that are primarily intended for liquid chromatographic columns. These resins are also useful for SPE. The PRP-1 and PRP-3 resins from Hamilton, PLRP-S from Polymer Labs, and MP-DVB from Interaction are examples of this type. However, chromatographic columns packed with 5–10-μm resins have a substantial backpressure. A SPE column packed with particles of this small size is feasible if the resin bed height is only a few millimeters. Careful packing and design of such a column is essential to avoid channeling through the very short resin bed.

Several sorbents with special properties are listed in Table 3.6. MP-1 (Interaction) contains C18 alkyl functional groups, which makes this material more hydrophobic than ordinary crosslinked polystyrene polymers. At the other end of the adsorption spectrum, Oasis (Waters) is easily wettable

**TABLE 3.6 Polymeric Sorbents for SPE**

| Supplier/Sorbent | Type | Surface Area, $m^2/g$ | Pore Size, Å | Particle Size |
|---|---|---|---|---|
| Hamilton | | | | |
| PRP-1 | PS-DVB | — | 100 | 10 μm |
| PRP-3 | PS-DVB | — | 300 | 10 μm |
| PRP-infinity | PS-DVB | — | Nonporous | 10 μμ |
| Interaction | | | | |
| MP-1 | PS-DVB + C18 groups | — | — | 20–60 μμ |
| MP-2 | Polyvinyl pyridine | — | — | 20–60 μm |
| MP-DVB | PS-DVB | — | — | 20–60 μm |
| Lab instruments | | | | |
| Ostion SP-1 | PS-DVB | 350 | 85 | — |
| Synachrom | PS-DVB-EVB | 520–620 | 90 | — |
| Spheron MD | Methacrylate-DVB | 320 | — | — |
| Polymer Labs | | | | |
| PLRP-S | PS-DVB | — | — | 10 μm |
| Supelco | | | | |
| XAD-2 | PS-DVB | 300 | 90 | 20–60 mesh |
| CG-161 | PS-DVB | 900 | 150 | 80–160 μm |
| CG-71 | Polymethacrylate | 500 | 250 | 80–160 μm |
| DAX-8 | Polymethacrylate | 160 | 225 | 40–60 mesh |
| CG-300s | PS-DVB | 700 | 300 | 20–50 μm |
| CG-300m | PS-DVB | 700 | 300 | 50–100 μm |
| CG-1000s | PS-DVB | 250 | 1000 | 20–50 μm |
| Tessek (Czeck Rep.) | | | | |
| Seperon HEMA | Hydroxyethyl-methacrylate | 20–60 | — | — |
| Waters | | | | |
| Oasis | Poly-DVB-vinylpyrrolidone | 830 | 82 | — |

by water alone and is able to take up many hydrophilic compounds efficiently. Oasis may be used for serum and other biological samples without any sample pretreatment.

For the most part crosslinked polystyrene resins have been the most successful for solid-phase extraction. Polyacrylates such as Rohm and Haas XAD-7 and XAD-8 are more polar than polystyrene particles and have been found to be superior for uptake of polar compounds such as fulvic acid (14) or phenol (15).

Spheron SE (16) and Separon SE (17) have been found suitable for preconcentration of toluene, cresols, phenoxycarboxylic acids, and *S*-triazines. The ability of Separon HEMA (hydroxyethylmethacrylate) to retain solutes would appear to be limited by its rather low surface area (20–60 $m^2/g$).

Tenax GC, poly(2,6-diphenyl-*p*-phenyleneoxide), is a very popular material for determining organic analytes in their vapor form in purge-and-trap procedures (18–22). In this method air or another gas is bubbled through an aqueous sample causing volatile compounds to be vaporized. The vapors are then trapped on a short column of Tenax resin. The success of this method lies in the fact that Tenax retains organic vapors well but has almost no affinity for water vapor. The unusually high temperature stability of Tenax (up to 380°C) and high purity permit adsorbed compounds to be eluted simply by heating the column. Often the adsorbed analytes are desorbed thermally directly into a gas chromatograph for analysis.

Tenax has also found some use for accumulating organic solutes from aqueous samples (23–26). Preconcentration efficiency has been found to be best for nonpolar compounds with low solubility in water such as polychlorinated biphenyls (27), polycyclic aromatic hydrocarbons (26,28), and various pesticides (26,28). However, the low surface area of Tenax 20 $m^2/g$ can lead to low recoveries of analytes from water (24). Another drawback of Tenax is its low resistance to organic solvents that might be used to elute adsorbed analytes. Some organic solvents will actually dissolve the Tenax.

### 3.4.1 Modified Polymeric Sorbents

Bonded-phase silica materials are readily available with a variety of hydrophobic (C18, C8, C2, etc.) and hydrophilic substituents (cyano, diol, amino, etc.). The sorptive properties of crosslinked polystyrene resins have also been modified by the introduction of various functional groups (29). The modified resins are easily prepared by a Friedel–Crafts reaction with the benzene ring of the polymer. Examples are (29)

*–$CH_2OH$ Derivatives.* React with formaldehyde, acetic acid, acetic anhydride, and anhydrous zinc chloride to form –$CH_2OCOCH_3$, then hydrolyze with methanol–hydrochloric acid.

*$-COCH_2CH_2CO_2H$ Derivatives.* React with succinic anhydride and aluminum chloride in tetrachloroethane–nitrobenzene.

*$-COCH_3$ Derivatives.* React with acetyl chloride and aluminum chloride in $CS_2$.

*$-C(CH_3)_3$ Derivatives.* React with *tert*-butyl chloride and aluminum chloride in nitrobenzene solution.

*$-CH_2CH$ Derivatives.* React with chloroacetonitrile and aluminum chloride in tetrachloroethane.

These resins have been used for HPLC as well as for SPE. Capacity factors for several organic analytes are compared in Table 3.7 with 50% aqueous acetonitrile as the mobile phase. Introduction of nonpolar *tert*-butyl groups leads to higher capacity factors for the more hydrophobic analytes (cumene, toluene, etc.). However, the capacity factors for these analytes are all lower on the resins substituted with polar groups ($-CH_2OH$, $-COCH_3$, etc.). The capacity factors for the most polar analytes (*p*-cresol and phenol) are higher on these resins, particularly on resins containing an acetyl group.

The effect of various functional groups on capacity factor can be seen from the data on Table 3.8, where the ratio of capacity factors of the substituted benzene to benzene itself are compared on three PS-DVB resins: underivatized, acetyl, and *tert*-butyl. The ratio ($R$) of the capacity factor on the substituted resin to that on the unsubstituted resin is also given for each analyte.

Solid-phase extraction of organic analytes is usually carried out with primarily aqueous samples. The capacity factors given above for 50% acetonitrile are useful in predicting the behavior of analytes with various resins in aqueous solution. The capacity factors increase considerably as the percentage of acetonitrile in the aqueous solution is decreased, but the order of $k'$ values of the various analytes tends to remain the same. In the middle range of organic–water solvent composition (~20–80% organic), log $k$ increases nearly linearly with decreasing percentage of organic in the liquid phase. The increase of log $k$ with decreasing organic content often becomes steeper as the liquid composition approaches that of pure water. This effect is shown in Figure 3.1 (PS-DVB) and Figure 3.2 (acetyl PS-DVB). It will be noted that capacity factors of the more polar analytes approach a higher value with the acetyl resin. Some of the $k$ values were too high to measure in water alone.

For comparison a plot of log $k$ versus percent ACN with C18 silica particles is given in Figure 3.3. In this case the change in log $k$ between 0% and 10% acetonitrile is much less than with the two PS-DVB materials. The

**TABLE 3.7 Capacity Factors ($k$) of Test Compounds on XAD-4 Resin and Its Derivatives[a]**

| | | Derivatives of XAD-4 Resin | | | | | | | | | |
|---|---|---|---|---|---|---|---|---|---|---|---|
| | XAD-4, | $-C(CH_3)_3$ | | $-CH_2OH$ | | $-COCH_3$ | | $-COCH_2-CH_2COOH$ | | $-CH_2CN$ | |
| Compound | $k$ | $k$ | $R$ | $k$ | $R$ | $k$ | $R$ | $k$ | $R$ | $k$ | $R$ |
| Cumene | 19.3 | 23.2 | 1.20 | 18.4 | 0.80 | 14.2 | 0.74 | 17.7 | 0.92 | 15.6 | 0.81 |
| *o*-Dichlorobenzene | 18.7 | 22.0 | 1.18 | 17.0 | 0.91 | 16.9 | 0.91 | 18.7 | 1.00 | 17.1 | 0.91 |
| Toluene | 9.83 | 11.5 | 1.17 | 8.83 | 0.90 | 8.50 | 0.86 | 8.83 | 0.90 | 8.78 | 0.89 |
| Anisol | 6.53 | 7.33 | 1.12 | 5.81 | 0.89 | 6.22 | 0.95 | 5.83 | 0.89 | 5.67 | 0.87 |
| Diethylphthalate | 6.25 | 7.30 | 1.19 | 4.83 | 0.77 | 4.58 | 0.73 | 5.17 | 0.83 | 3.89 | 0.78 |
| Methylbenzoate | 5.25 | 5.83 | 1.11 | 4.84 | 0.92 | 4.61 | 0.88 | 4.15 | 0.90 | 4.78 | 0.91 |
| 2,4-Dinitrofluorobenzene | 4.92 | 4.41 | 0.89 | 4.50 | 0.92 | 4.28 | 0.87 | 4.83 | 0.98 | 4.08 | 0.83 |
| Acetophenone | 3.17 | 3.25 | 1.03 | 3.17 | 1.00 | 3.00 | 0.95 | 3.00 | 0.95 | 3.00 | 0.95 |
| *p*-Cresol | 2.17 | 1.92 | 0.88 | 2.25 | 1.04 | 2.69 | 1.24 | 2.08 | 0.96 | 2.11 | 0.97 |
| Phenol | 1.50 | 1.42 | 0.94 | 1.67 | 1.11 | 1.94 | 1.30 | 1.67 | 1.11 | 1.61 | 1.07 |

[a] R is the ratio of $k$ on the derivatized resin to $k$ on the XAD-4 resin. Chromatographic conditions: $250 \times 2.1 =$ mm column; acetonitrile–water (50–50) eluant, adjusted to pH 1.7 with HCl.

*Source:* From Reference 9 with permission.

**TABLE 3.8 Capacity Factors ($k$) and Capacity Factors Relative to $k$ (Benzene) for Benzene Derivatives on Three Polymeric Resin Columns**

| | PS-DVB | | PS-DVB-$COCH_3$ | | | PS-DVB-$C(CH_3)_3$ | | |
|---|---|---|---|---|---|---|---|---|
| Compounds | $k$ | $k_a/k_{benzenes}$ | $k$ | $k_a/k_{benzenes}$ | $R$ | $k$ | $k_a/k_{benzenes}$ | $R$ |
| Benzene | 6.59 | — | 4.82 | — | — | 7.21 | — | — |
| Biphenyl | 41.28 | 6.27 | 28.46 | 5.90 | 0.69 | 50.63 | 7.02 | 1.23 |
| Cumene | 22.98 | 3.49 | 11.25 | 2.33 | 0.49 | 28.71 | 3.98 | 1.25 |
| *o*-Dichlorobenzene | 19.18 | 2.91 | 12.38 | 2.57 | 0.65 | 24.29 | 3.37 | 1.27 |
| Iodobenzene | 17.89 | 2.71 | 13.83 | 2.87 | 0.77 | 23.19 | 3.22 | 1.30 |
| Bromobenzene | 15.53 | 2.36 | 12.93 | 2.68 | 0.83 | 19.78 | 2.74 | 1.27 |
| Chlorobenzene | 11.98 | 1.82 | 10.18 | 2.11 | 0.85 | 15.05 | 2.09 | 1.26 |
| Toluene | 10.83 | 1.64 | 6.61 | 1.37 | 0.61 | 12.84 | 1.78 | 1.19 |
| Anisol | 6.93 | 1.05 | 4.50 | 0.93 | 0.65 | 7.73 | 1.07 | 1.12 |
| Diethylphthalate | 6.33 | 0.96 | 2.97 | 0.62 | 0.47 | 6.72 | 0.94 | 1.06 |

| | | | | | | | | |
|---|---|---|---|---|---|---|---|---|
| Fluorobenzene | 5.99 | 0.91 | 5.17 | 1.07 | 0.86 | 6.85 | 0.95 | 1.14 |
| Methylbenzoate | 5.48 | 0.79 | 3.57 | 0.74 | 0.65 | 6.12 | 0.85 | 1.12 |
| Nitrobenzene | 5.24 | 0.80 | 4.70 | 0.98 | 0.90 | 5.99 | 0.83 | 1.14 |
| 2,4-Dinitrofluorobenzene | 4.56 | 0.69 | 4.08 | 0.85 | 0.89 | 4.29 | 0.60 | 0.94 |
| Acetophenone | 3.06 | 0.46 | 2.05 | 0.43 | 0.67 | 3.25 | 0.45 | 1.06 |
| *p*-Cresol | 1.70 | 0.26 | 1.96 | 0.41 | 1.15 | 1.65 | 0.23 | 0.97 |
| 3-Nitrophenol | 1.59 | 0.24 | 2.07 | 0.43 | 1.30 | 1.50 | 0.21 | 0.94 |
| 4-Nitrophenol | 1.43 | 0.22 | 1.97 | 0.41 | 1.38 | 1.27 | 0.18 | 0.89 |
| Phenol | 1.23 | 0.19 | 1.42 | 0.29 | 1.15 | 1.11 | 0.15 | 0.90 |
| Benzoic acid | 0.78 | 0.12 | 2.55 | 0.53 | 3.27 | 0.75 | 0.10 | 0.96 |
| Benzyl alcohol | 0.76 | 0.12 | 0.93 | 0.19 | 1.22 | 0.71 | 0.10 | 0.93 |

[a] $R$ is the ratio of $k$ on the derivatized resin compared to the underivatized resin. The eluant was acetonitrile–water (50–50), acidified with HCl. $k_a$ is the capacity factor of a benzene derivative.

*Source:* From Reference 5 with permission.

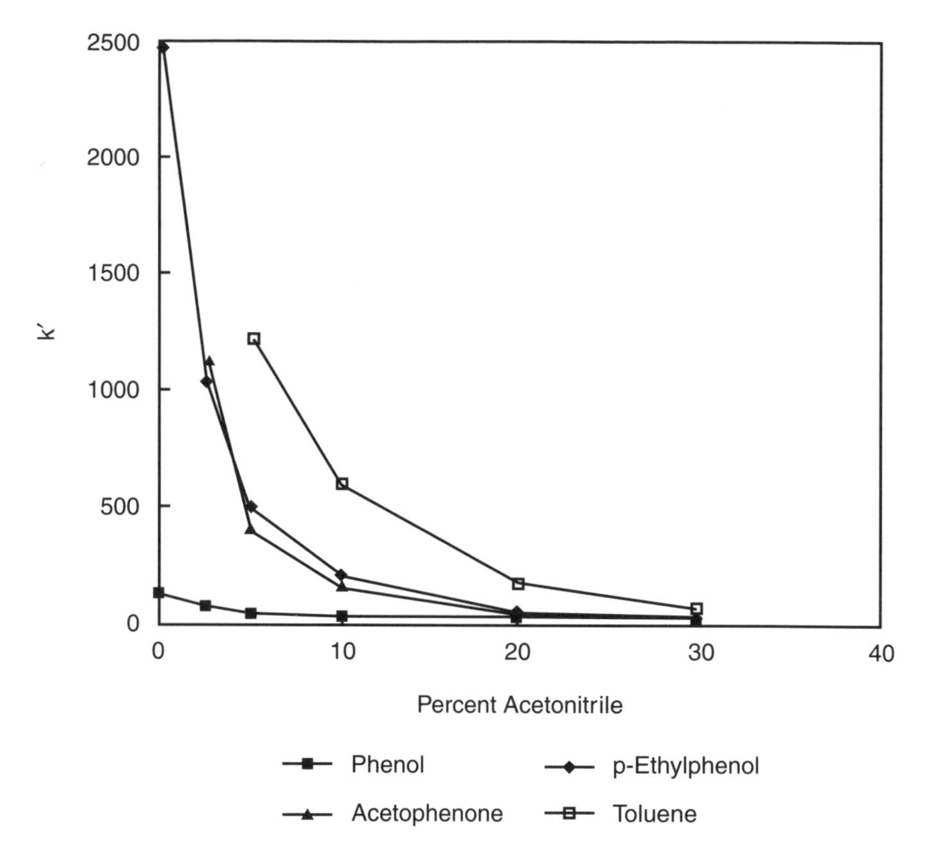


**FIGURE 3.1** Capacity factors on PS-DVB resin as a function of acetonitrile percentage in the organic–aqueous mobile phase.

values of log $k$ are also significantly lower with the silica sorbent for each of the analytes. This behavior may help explain the observation that in SPE the percentage recoveries of analytes often tend to be higher with PS-DVB resins than with C18 silica.

In SPE intimate interfacial surface contact between the aqueous sample and the resin is essential for efficient extraction of organic analytes. The bulk of the solid particles must be hydrophobic to extract the analytes effectively, but the solid particle surfaces need to be somewhat hydrophilic for proper interfacial contact. Octadecylsiloxane and PS-DVB sorbents both present some problems in that their surfaces are too hydrophobic to be effectively wetted by water alone. The usual answer to this problem is to pretreat the SPE column with a water-miscible organic solvent such as

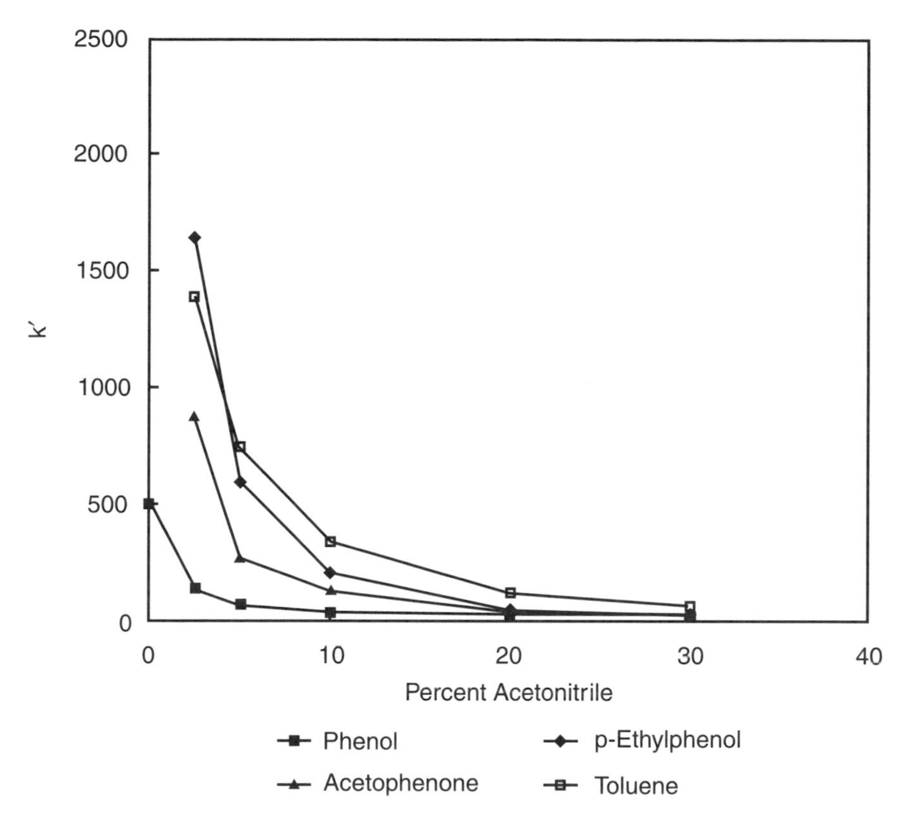


**FIGURE 3.2** Capacity factors on acetyl PS-DVB resin.

methanol. Some of the methanol is adsorbed, thus giving a more wettable surface. But the organic solvent can be gradually lost from the sorbent, particularly if the liquid level falls below the top of the resin bed, thus allowing air to be sucked into the column.

A better answer to the surface contact problem is to use a PS-DVB resin with hydrophilic groups on its surfaces. Sun and Fritz found acetyl- and hydroxymethyl-substituted PS-DVB to be the most effective in this regard (5). They compared C18 silica, underivatized Amberchrome (a PS-DVB resin from Rohm and Haas), hydroxymethol Amberchrome, and acetyl Amberchrome for SPE of a number of organic analytes. In each case the extractions were done both with and without the usual methanol pretreatment. The results in Table 3.9 show higher average recoveries of the test compounds with the derivatized for the "wet" runs (methanol pretreatment). All the PS-DVB resins gave higher average recoveries than did the C18

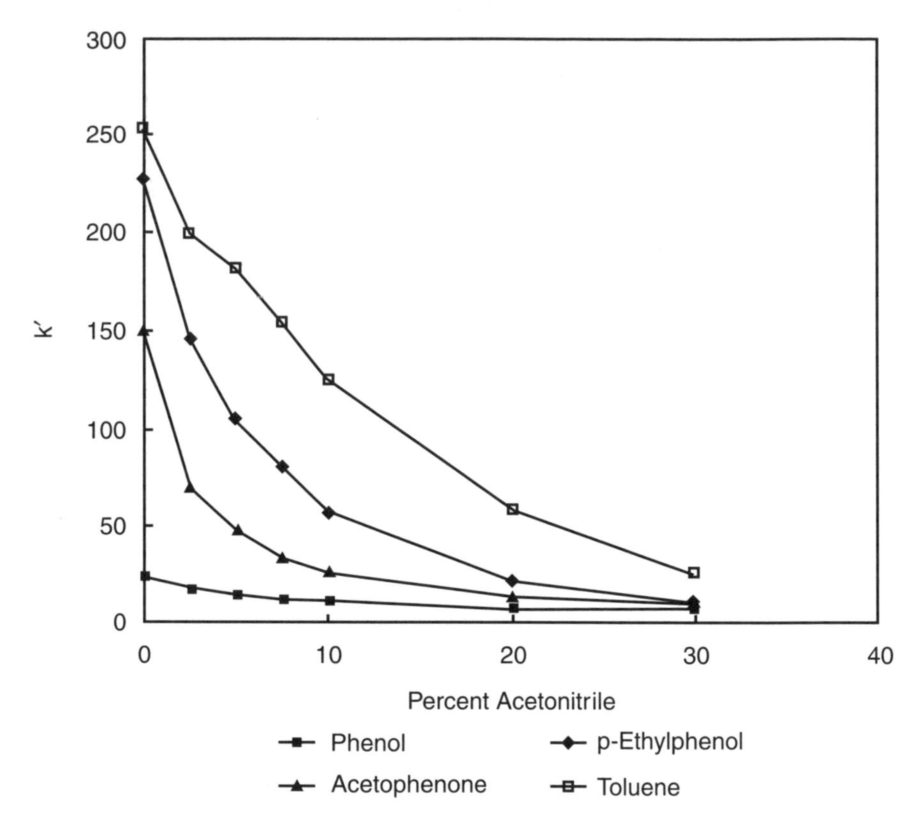


**FIGURE 3.3** Capacity factors on a ODS silica column.

silica materials. For the "dry" runs (no methanol pretreatment) the differences were even greater. Here, recoveries were unacceptably low except for the two derivatized resins where the average recoveries were only about 10% lower than those obtained with methanol pretreatment.

Other results reported by Sun and Fritz (5) and by continuing studies in the author's laboratory confirm the following conclusions:

1. Some test compounds give just as high recoveries on C18 silica as on PS-DVB materials, but in general recoveries are significantly higher on the polymeric sorbents.
2. On the average, recoveries of test compounds are better on resins with hydrophilic- surface substituent than with the parent, unfunctionalized resin.

**TABLE 3.9 Recoveries of Phenols, Aromatic Compounds, and Polyhydroxy Aromatic Compounds by SPE under Wet and Dry Loading Conditions**

| | Recovery (%) | | | | | | | |
|---|---|---|---|---|---|---|---|---|
| | C18Si | | Amberchrome | | Amberchrome–$CH_2OH$ | | Amberchrome–$COCH_2$ | |
| Compound | Wet | Dry | Wet | Dry | Wet | Dry | Wet | Dry |
| Phenol | 6 | <3 | 91 | 3 | 94 | 75 | 100 | 93 |
| *p*-Cresol | 16 | 4 | 91 | 12 | 98 | 88 | 101 | 94 |
| *p*-Ethylphenol | 66 | 15 | 96 | 37 | 99 | 97 | 101 | 99 |
| 2-Nitrophenol | 45 | 17 | 93 | 47 | 95 | 96 | 96 | 96 |
| 3-Nitrophenol | <5 | <5 | 81 | <5 | 85 | 75 | 93 | 73 |
| 4-Nitrophenol | <5 | <5 | 87 | <5 | 86 | 77 | 87 | 85 |
| 2,4-Dimethylphenol | 71 | 21 | 95 | 42 | 97 | 96 | 100 | 98 |
| 4-*tert*-Butylphenol | 83 | 49 | 88 | 50 | 96 | 90 | 100 | 95 |
| Anisol | 78 | 58 | 91 | 56 | 94 | 95 | 98 | 96 |
| Aniline | 9 | <5 | 94 | 26 | 96 | 90 | 100 | 96 |
| Benzylalcohol | 10 | <5 | 92 | 17 | 98 | 85 | 99 | 99 |

*(cont.)*

**TABLE 3.9** *(Continued)*

| | Recovery (%) | | | | | | | |
|---|---|---|---|---|---|---|---|---|
| | C18Si | | Amberchrome | | Amberchrome–$CH_2OH$ | | Amberchrome–$COCH_2$ | |
| Compound | Wet | Dry | Wet | Dry | Wet | Dry | Wet | Dry |
| Nitrobenzene | 54 | 27 | 92 | 51 | 96 | 96 | 100 | 97 |
| 2,4-Dinitrofluorobenzene | 44 | 4 | 83 | 23 | 96 | 92 | 98 | 94 |
| *o*-Hydroxyacetophenone | 88 | 67 | 85 | 54 | 95 | 94 | 96 | 94 |
| Isopentylbenzoate | 84 | 60 | 72 | 73 | 89 | 84 | 96 | 85 |
| Diethylphthalate | 90 | 70 | 87 | 58 | 96 | 84 | 100 | 90 |
| Average | 47 | 26 | 89 | 35 | 94 | 88 | 98 | 93 |
| Catechol | 0 | 0 | 72 | 0 | 89 | 9 | 75 | 30 |
| Resorcinol | 0 | 0 | 61 | 0 | 88 | 0 | 97 | 95 |
| *o*-Methylresorcinol | 0 | 0 | 83 | 0 | 97 | 16 | 99 | 96 |
| Hydroquinone | 0 | 0 | 26 | 0 | 72 | 0 | 87 | 81 |
| Methylhydroquinone | 0 | 0 | 77 | 0 | 98 | 5 | 99 | 94 |
| Phloroglucinol | 0 | 0 | 0 | 0 | 24 | 0 | 56 | 42 |

*Source:* From Reference 9 with permission.

Grienberger (30) described the preparation and use of a PS-DVB resin with pendant C18 groups. This material was found to give an average recovery of 98% for solid-phase extraction of 13 pesticides. By comparison, the average recovery with C18 silica was 89%.

The resin was prepared by a two-step polymerization that gave particles that were extremely uniform in size. The first step is an emulsion polymerization that gave particles approximate 1 μm in diameter. The second step consists of treating the primary particles with a swelling reagent (SDS and 1-chlorodecane). In the third step styrene and divinylbenzene are added to the suspension of swollen particles together with toluene as a porogen. These substances penetrate into the swollen particles, causing them to increase in size. When the particles have attained the desired size as determined by a light microscope, the strongly swollen particles are separated from the excess of the monomers.

In the fourth and last step the monomers inside the swollen particles are copolymerized by addition of dibenzylperoxide as a radical initiator and by maintaining a temperature of 70 °C. After washing the resin particles, C18 groups arc introduced by a Friedel–Crafts reaction with 1-chlorooctadecane and aluminum chloride catalyst.

## 3.5 CARBONS

It should be recognized that there are many different types of carbon sorbents and that their properties can vary considerably. Carbonaceous materials for SPE range from activated carbon, which was popular in the 1970s, to more modern graphitized carbon sorbents. The high surface area and strong sorptive properties of carbon have contributed to their long-standing popularity. But elution of adsorbed materials from carbon is often slow and incomplete. The catalytic activity of a carbon surface can also be a problem (31–33).

Chriswell et al. (34) carried out extensive comparisons of activated carbon and a popular polymeric resin, Rohm and Haas XAD-4, for SPE. The percentage recoveries of the compounds studied at 100 μg/liter were generally lower for the activated carbon. Since the values given are for *overall* recoveries, the less-than-complete recoveries could be due to (1) incomplete adsorption from the aqueous sample, (2) incomplete elution by 1:1 acetone–chloroform, or (3) both of the above. Accordingly, each of the analytes was tested for in the aqueous effluent from the SPE column. Although there were some instances in which the test compound was incompletely extracted, the major reason for low recovery seemed to be incomplete elution.

The comparative recoveries of the 100 compounds studied in Table 3.10 can be summarized as follows: use of the XAD-4 resin column resulted in a higher recovery than with carbon for 59 of the compounds while recovery of only 12 of the compounds was higher on the activated carbon. Comparable recoveries were obtained for 7 of the compounds, and neither column was adequate for 22 of the test compounds. The recoveries of the test compounds according to class are given in Table 3.10. The resin was more effective for 9 classes, including phthalate esters, most aromatic compounds, and pesticides. The carbon column was more effective for alkanes, while neither sorbent was very good for acidic compounds.

A product known as *Spherocarb* became available during the late 1970s (35). It is supplied as 100–200-mesh spherical particles and has a surface area of approximately 1200 $m^2/g$. It is strong mechanically and is classified as a molecular sieve. Tateda and Fritz (35) compared Spherocarb to a PS-DVB material (Rohm and Haas XAD-4) for SPE. After the sorption step, acetone, acetonitrile, diethyl ether, methanol, methylenechloride, pentane, and carbon disulfide were evaluated as eluting solvents. In general, the

**TABLE 3.10 Comparative Recovery of Classes of Test Compounds Using XAD-4 Resin and Activated Carbon**

| | Overall Recovery, % | |
|---|---|---|
| Types of Compound (Number) | Resin | Activated Carbon |
| Alkanes (5) | 5 | 15 |
| Esters (4) | 61 | 49 |
| Alcohols (8) | 73 | 47 |
| Phthalate esters (3) | 82 | 24 |
| Phenols (10) | 45 | 7 |
| Chlorinated alkanes and alkenes (5) | 43 | 55 |
| Chlorinated aromatic compounds (13) | 70 | 11 |
| Aromatic compounds (7) | 68 | 6 |
| Aldehydes and ketones (3) | 74 | 4 |
| Amines (13) | 54 | 24 |
| Carboxylic acids (11) | 1 | 2 |
| Pesticides (4) | 34 | 16 |
| Miscellaneous (14) | 33 | 11 |

*Source:* Data from Reference 34.

elution process was found to be very slow and a relatively large volume of organic solvent was required for complete desorption of the analytes. However, carbon disulfide was found to be satisfactory for elution of most compounds.

In the comparative study, the columns used were 25 × 1.2 mm or 25 × 1.8 mm. Only 50–100 μl of an organic solvent was needed for effective elution of the retained test compounds. The results obtained for 17 test compounds are given in Table 3.11. Omitting 1-butanol, the average recovery was 81% on Spherocarb and 82% on XAD-4. The recovery of diethyl malonate was 0% on the carbon but was 87% on the resin. On the other hand, recovery of 1-pentanol was 67% on carbon but only 9% on XAD-4.

**TABLE 3.11 Comparison of Recoveries of Test Compounds at 100-ppb Level from Spherocarb and XAD-4 Resin**

| | Recovery, % | | | |
|---|---|---|---|---|
| | Spherocarb | | XAD-4 | |
| Compound | Water | 10% Methanol | Water | 10% Methanol |
| 1-Butanol | 10 | 0 | 0 | 0 |
| 1-Pentanol | 67 | 66 | 9 | 6 |
| 1-Hexanol | 81 | 83 | 81 | 47 |
| 1-Octanol | 85 | 90 | 85 | 99 |
| 1-Decanol | 90 | 92 | 85 | 90 |
| Chlorobenzene | 87 | 82 | 90 | 78 |
| *o*-Dichlorobenzene | 89 | 88 | 90 | 84 |
| 1,2,4-Trichlorobenzene | 90 | 87 | 80 | 80 |
| *p*-Chlorotoluene | 86 | 83 | 86 | 83 |
| 2-Heptanone | 83 | 87 | 89 | 89 |
| 2-Octanone | 85 | 92 | 88 | 88 |
| 2-Nonanone | 87 | 93 | 87 | 86 |
| Methylhexanoate | 88 | 90 | 96 | 88 |
| Methyloctanoate | 87 | 91 | 95 | 89 |
| Diethylmalonate | 0 | 0 | 87 | 90 |
| Methyldecanoate | 89 | 89 | 96 | 83 |
| Methylbenzoate | 98 | 94 | 100 | 97 |

*Source:* Data from Reference 35.

Addition of 10% v/v methanol to the aqueous test sample had almost no effect on the percentage recoveries.

### 3.5.1 Porous Graphitic Carbon

Graphitized carbon black has been used successfully for concentration of chloroanilines, chlorophenols, and moderately polar pesticides, but its poor mechanical properties hinder its use in chromatographic columns (36–39).

The properties of a pyrolyzed graphitic carbon known as *Hypercarb* (Shandon, Runcorn, UK) became available in the 1980s. Its physical and chemical properties have been studied in detail (40). The authors concluded that Hypercarb is a strong hydrophobic adsorbent. It has favorable physical properties and is available in 7-μm spherical particles. Because of its strong adsorptive properties, a higher proportion of organic solvent is needed in the organic–aqueous mobile phase used in liquid chromatography; typically, 95% methanol is used with Hypercarb compared with ~50% methanol with BP silica materials.

Coquart and Hennion used Hypercarb for trace enrichment of polar phenolic compounds (41). Pyrocatechol, resorcinol, and phloroglucinol were successfully concentrated at < 0.1 μg/liter from 50-ml aqueous samples. Polar compounds such as these normally would be retained very poorly by conventional SPE materials.

Supelco (Bellefonte, PA, USA) offers ENVI-Carb, a nonporous graphitic carbon, 120–140 mesh, with a surface area of 100 $m^2/g$.

## 3.6 COMPARISON OF SORBENTS

In the previous sections some general comparisons were made concerning the percentage recoveries of various solid extractants. For example, it was noted that recoveries of test compounds were often significantly higher with polymeric organic particles than with bonded-phase silicas. The relative extractive power of the various solid-phase extractants can perhaps best be compared by measuring their retention factors ($k$) with analytes of different chemical structures. Because these $k$ values are often too high to measure in water alone, Hennion and Coquart (42) measured retention factors in a series of methanol–water solutions and obtained the values in water alone ($k_w$) by linear extrapolation. Some error is involved in these extrapolations, but the $k_w$ values obtained are good enough for general comparisons.

The $k_w$ values in Table 3.12 were selected from this reference. The following conclusions may be drawn from these and other data.

1. The $k_w$ values of moderately polar compounds (aniline, benzoic acid, phenol, etc.) are highest with PRP-1 (a crosslinked polystyrene) and lowest with silica C18.
2. The $k_w$ values of highly polar compounds (*p*-aminophenol, polyhydroxyphenols) are high enough for effective SPE only with PGC.
3. The $k_w$ values of more hydrophobic compounds are higher with PRP-1 than with C18 silica.

It is not always necessary to choose the sorbent that gives the highest $k$ values. The purity and wide availability of BP silica materials continue to make them attractive so long as the estimated $k$ value is sufficient ($k \cong 100$) to give consistent and reasonably high recoveries. Polymeric materials, such as PRP-1, are the particles of choice when a high concentration factor is needed or when the $k$ for BP silica is too low. PGC is indicated for samples containing very polar analytes.

**TABLE 3.12 Retention Factors in Water ($k_w$) for C18 Silica, PRP-1, and Pyrolytic Graphitized Carbon (PGC)**

| | $k_w$ | | |
|---|---|---|---|
| Compound | C18 Silica | PRP-1 | PGC |
| Acetophenone | 63 | 1,250 | — |
| 4-Aminophenol | — | 13 | 110 |
| Aniline | 12 | 200 | 22 |
| Benzaldehyde | 55 | 875 | — |
| Benzene | 160 | 3,300 | 28 |
| Benzoic acid | 80 | 2,000 | 250 |
| Benzylalcohol | 25 | 280 | — |
| 1,4-Dihydroxybenzene | — | 7 | 140 |
| Ethylbenzene | 2,500 | 63,000 | — |
| Nitrobenzene | 110 | 4,200 | — |
| 4-Nitrophenol | 70 | 630 | — |
| Phenol | 35 | 250 | 63 |
| Toluene | 560 | 14,000 | — |
| Trihydroxybenzene | — | 3 | 500 |

## 3.7 MOLECULAR SEIVES

A class of compounds known as *molecular sieves* has found numerous uses in chemical analysis. Molecular sieves are silica or metal–silica particles that contain cavities that are usually 5–6 Å in diameter. Compounds small enough to enter the cavities are sorbed by the molecular sieve while larger molecules are excluded. Virtually all molecular sieves are hydrophilic.

A molecular sieve known as *Silicalite*, first synthesized in 1977 (43) and now available from UOP, has the rather unique property of being hydrophobic. Most molecular sieves are strongly hydrophilic and can be used to sorb water from organic solvents and gas streams. Because it is hydrophobic, Silicalite can be used as a sorbent for various organic and inorganic solutes from liquid and gaseous samples (44,45). It has also been used to remove ethanol from beer (46).

Silicalite is a polymorph of silica, with a rather unusual crystal structure. Its tetrahedral framework consists largely of five-membered rings of silicon–oxygen tetrahedra. Silicalite has intersecting channels defined by rings of 10 oxygen atoms. These channels are 6 Å in diameter. The total pore volume of Silicalite is approximately 33% (44).

Chromatographic applications of Silicalite have been rare. Schultz-Sibbel et al. listed distribution ratios of a number of solutes between gas or water and Silicalite (44). Fritz and Ogawa used Silicalite as an adsorbent for solid-phase extraction of aldehydes and ketones from aqueous samples (45). Andronikashvili et al. used Silicalite as a stationary phase in gas chromatography (47). Dumont obtained excellent chromatographic separations of substituted phenols and several cis-trans isomers on columns packed with Silicalite (48).

A major drawback of bonded-phase silicas and crosslinked PS-DVB resins is that these materials do not extract, or extract very poorly, small, polar organic compounds such as the lower alcohols, aldehydes, amines, carboxylic acids, and ketones. However, many of these polar compounds are extracted very well by small tubes filled with Silicalite particles (49). Table 3.13 compares the recoveries of test compounds extracted by Silicalite and by a membrane disk containing an efficient polystyrene sorbent. In several instances the lower members of a homologous series are retained much more strongly by Silicalite than by the polymeric material.

The effect of molecular size and shape on SPE was studied by comparing the SPE recoveries of alcohols by Silicalite and by a lightly sulfonated PS-DVB material (49). The percentage recoveries of the normal (straight-chain) alcohols are plotted as a function of carbon number in Figure 3.4.

**TABLE 3.13 Comparison of Percentage Recoveries of Various Test Compounds from Aqueous Solutions Using Silicalite Particles and a Sulfonated PS-DVB Resin-Loaded Membrane**

| Class | Compounds | Silicalite | Membrane |
|---|---|---|---|
| Aldehydes | *trans*-Crotonaldehyde | 91 | 24 |
| | *n*-Valeraldehyde | 100 | 91 |
| | Hexanal | 107 | 85 |
| | Nonylaldehyde | 84 | 74 |
| | Benzaldehyde | 89 | 94 |
| | Salicylaldehyde | 54 | 96 |
| Ketones | Acetones | 94 | 0 |
| | 2-Butanone | 92 | 0 |
| | 2-Pentanone | 90 | 88 |
| | 3-Pentanone | 41 | 68 |
| | 4-Methyl-2-pentanone | 104 | 88 |
| | 2,4-Pentadione | 31 | 27 |
| | 2-Hexanone | 93 | 89 |
| | 3-Hexanone | 81 | 89 |
| Esters | Methylformate | 74 | 0 |
| | Methylacetate | 85 | 4 |
| | Ethylformate | 83 | 0 |
| | Ethylacetate | 91 | 55 |
| | Ethylpropionate | 88 | 61 |
| | Ethylbutyrate | 90 | 75 |
| | Hexylacetate | 79 | 88 |
| | Methylbenzoate | 68 | 94 |
| | Pentylbenzoate | 82 | 70 |
| Chlorinated alkanes | Chloroform | 82 | 81 |
| | 1,2-Dichlorethane | 83 | 77 |
| | 1,1-Dichlorethane | 80 | 77 |
| | 1,2-Dichloropropane | 85 | 85 |
| Carboxylic acids | Acetic | 2 | 0 |
| | Propionic | 62 | 0 |
| | Butyric | 78 | 13 |
| | Valerie | 95 | 104 |
| Amines | Propylamine | 63 | 0 |
| | *n*-Butylamine | 72 | 52 |
| | *tert*-Butylamine | 64 | 84 |
| | Pentylamine | 78 | 93 |
| | Pyridine | 77 | 35 |
| | Tributylamine | 13 | 66 |

*Source:* From Reference 49 with permission.

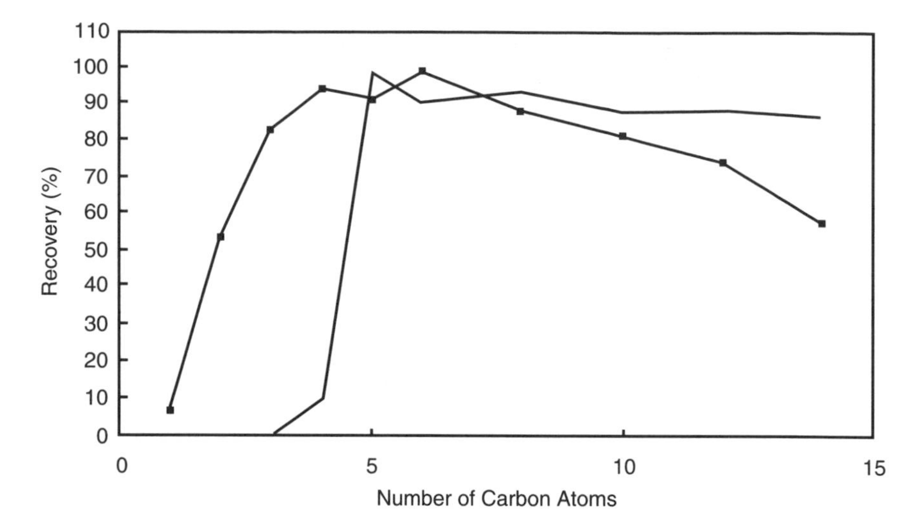


**FIGURE 3.4** SPE percentage recoveries of *n*-alcohols using silicalite particles (■) and a sulfonated PS-DVB resin-loaded membrane (solid line). (From Ref. 49 with permission.)

The most striking difference between the two sorbents is in the C1–C4 alcohols, which are much more strongly retained by Silicalite. From a maximum around C6, the Silicalite recoveries decrease gradually with increasing chain length. Recoveries using the sulfonated membrane remain almost constant between C5 and C14.

For alcohols with a *n*-alkyl group, increasing the chain length does not increase their effective diameter. The analytes can easily fit into Silicalite with their hydrocarbon chain parallel to the channels. The increased recoveries are thought to result from an increased interaction between paraffinic hydrogen atoms and the channel walls. As the hydrocarbon chain lengthens, extraction recoveries of the alcohols decrease. The longer hydrocarbon chain lessens the chance of an analyte directly entering the channels of Silicalite. An analyte will be sorbed if the energy needed to straighten the chain is less than the energy of sorption by Silicalite. Longer molecules need more energy to be straightened and, therefore, give lower extraction recoveries.

## 3.8 SORBENTS FOR NORMAL-PHASE SPE

Silica gels, activated alumina, and other inorganic adsorbents have found wide and varied uses in chemical analysis and chromatography. They retain hydrophilic organic material much more strongly than do compounds that are more hydrophobic. Normal-phase chromatography is carried out with a hydrophilic stationary phase and a mobile phase consisting of a very hydrophobic solvent such as hexane as a "carrier" with varying amounts of a more hydrophilic organic solvent as the "modifier."

Silica gels, and in particular Florisil (a magnesia silica gel), are used extensively in cleanup of samples prior to analysis. For example, in the analysis of vegetable matter for traces of herbicides or pesticides, an aqueous slurry of the sample is prepared in a blender. The pesticides are extracted from the slurry along with portions of the vegetable matter by a mixture of organic solvents that are not miscible with water. Then the organic extract is passed through a column containing Florisil (or another sorbent) and eluted with one or more organic solvents (or solvent mixtures). The pesticides are eluted while the more hydrophilic vegetable matter remains tightly held by the column. Some of the most used sorbents are listed in Table 3.14 together with their typical physical characteristics.

Alumina ($Al_2O_3$) is activated by heating to a high temperature to increase its sorptive properties and to render it insoluble in water. Alumina is neutral

**TABLE 3.14 Hydrophilic Adsorption Media**

| Type | Particle Size, μm | Pore Size, μm | Surface Area, $m^2/g$ |
|---|---|---|---|
| Alumina; acidic, basic, or neutral | 100 | 0.6 | 155 |
| Florisil (magnesium silica gel) | | | |
| Standard grade | 74–150 | — | 300 |
| PR grade | 150–250 | — | 300 |
| Hydroxylalkoxypropyl dextrans | 25–100 | — | — |
| C11–C14, C15–C18, or C18 | (dry) | — | — |
| Silica gels | | | |
| E. Merck | 65–200 | 0.4 | 750 |
| | 15–40, 40–65 | 0.6 | 550 |
| | 65–200 | 1.0 | 300 |
| Divisil | 35–75, 75–150 | 0.6 | 500 |
| | 75–150 | 1.5 | 300 |
| Spherosorb HP (hydroxypropyl dextran) | 10–25 | — | — |

at its isoelectric point (a pH a little above 8). At more acidic pH values alumina takes on a proton. This gives it a net positive charge and imparts anion exchange properties. At more acidic pH values, alumina takes on a proton. This gives it a net positive charge and imparts anion-exchange properties. At pH values above its isoelectric point, alumina acquires a negative charge and becomes a cation exchanger. Other than by ion exchange, alumina retains analytes by dipole–dipole interaction.

Florisil is a magnesia silica gel that can be activated by heating to 650°C. It is widely used to clean up liquid organic solvent extracts of plant and animal tissue. The PR grade is specifically activated for column cleanup and separation of chlorinated pesticides (50).

Hydroxylalkoxypropyl dextrans are prepared by incorporating different percentages of long-chain alkyl ethers into the hydroxypropyl dextran matrix. This yields a series of sorbents of varying and moderate hydrophilicity. One gram of the dry material swells to approximately 4 $cm^3$ when added to water or methanol. The gel structure necessitates the use of fairly gentle pressure or suction to achieve sample flow through a packed column.

Silica gel columns are used for normal-phase chromatography and for solid-phase extraction of polar analytes from hydrophobic organic solutions. A large number of silica gels are available with varying physical properties. Small differences in the amount of water in a silica gel can result in wide differences in ability to adsorb analytes from organic solutions.

Spherosorb HP is prepared by hydroxypropylation of a crosslinked dextran polymer. It has a small pore size and an intermediate polarity. The dry material swells appreciably to form a gel when added to water or methanol. One gram of the dry sorbent swells to a volume of approximately 2 $cm^3$.

Bonded-phase silica materials containing a highly polar functional group are also used for normal-phase SPE. Silica sorbents with cyanopropyl groups (silica–O–$CH_2CH_2CH_2CN$), diol groups (silica–O–$(CH_2)_3$-$OCH_2CH(OH)CH_2OH$), or aminopropyl groups (silica–O–$OCH_2CH_2CH_2NH_2$) are often used for this purpose.

## 3.9 OTHER SORBENTS

Resins having cation-exchange or anion-exchange properties also find major use in SPE. Cation exchangers include the so-called strong-acid type (-$SO_3$–$H^+$ groups) and the weak-acid type (–$CO_2$–$H^+$ groups). Strong-base anion exchangers contain quaternary ammonium groups (-$NR_3^+OH^-$), while weak-base resins contain an amino group that can take on a proton and become an anion exchanger. The types of ion-exchange particles available, together with principles and examples of their use in SPE, will be discussed in Chapter 5.

The practice of SPE is certainly not limited to the use of cartridges or minicolumns packed with sorbent particles. A growing family of resin-loaded membranes has become available. These membranes have some major advantages over the use of loose resins as extractants. SPE with resin-loaded membranes will be discussed in Chapter 6.

Before continuing further with a description of available sorbents, it seems appropriate to consider some practical aspects of apparatus and techniques used in solid-phase extraction. This will be done in the next chapter.

## REFERENCES


1. J. N. King and J. S. Fritz, *Anal. Chem.* **57** (1985) 1016.
2. G. A. Junk, in *Organic Pollutants in Water*, I. H. Suffet and M. Malayiandi, eds., ACS Symp Series 214, Washington, DC, 1987, p. 201.
3. G. A. Junk, M. J. Avery, and J. J. Richard, *Anal. Chem.* **60** (1988) 1347.
4. K. G. Miller and C. F. Poole, *J. HRC* **17** (1994) 125.
5. J. J. Sun and J. S. Fritz, *J. Chromatogr.* **522** (1990) 95–105.
6. C. H. Marvin, I. D. Brindle, C. D. Hall, and M. Chiba, *Anal. Chem.* **62** (1990) 1495.
7. G. A. Junk and J. J. Richard, *Anal. Chem.* **60** (1988) 451.
8. G. A. Junk, J. J. Richard, M. D. Grieser, D. Witiak, J. L. Witiak, M. D. Arguello, R. Vick, H. J. Svec, J. S. Fritz, and G. V. Calder, *J. Chromatogr.* **99** (1974) 745–762.
9. J. J. Sun and J. S. Fritz, *J. Chromatogr.* **590** (1992) 197.
10. J. J. Sun and J. S. Fritz, U.S. *Patent* 5, 071, 565 (Dec. 10, 1991).
11. L. Schmidt and J. S. Fritz, *J. Chromatogr.* **640** (1993) 145–149.
12. K. G. Miller and C. F. Poole, *J. HRC.* **17** (1994) 125.
13. I. Lŭska, J. Krupčik, and P. A. Leclercq, *J. High Resol. Chrom.* **12** (1989) 577.
14. G. R. Aiken, E. M. Thurman, R. L. Malcolm, and H. F. Walton, *Anal. Chem.* **51** (1979) 1799.
15. A. K. Burnham, G. V. Calder, J. S. Fritz, G. A. Junk, H. J. Svec, and R. Willis, *Anal. Chem.* **44** (1972) 139.
16. E. Brizová, M. Popl, and J. Coupek, *J. Chromatogr.* **139** (1977) 15.
17. Z. Voznáková, M. Popl, and M. Kovár, *Scientific Papers of the VSCHT*, Prague H19, 1984, p. 85.

18. T. A. Bellar, J. J. Lichtenberg, and R. C. Kroner, *J. Am. Waterworks Assoc.* **66** (1974) 739.
19. P. P. K. Kuo, E. S. K. Chian, F. B. De Walle, and J. H. Kim, *Anal. Chem.* **49** (1977) 1023.
20. W. Bertsch, E. Anderson, and G. Holzer, *J. Chromatogr.* **112** (1976) 701.
21. A. Zlatkis, W. Bertsch, H. A. Lichtenstein, A. Tishbee, F. Shunbo, N. M. Liebich, A. M. Coscia, and N. Flemschen, *Anal. Chem.* **45** (1973) 763.
22. J. M. Warner and R. K. Beasley, *Anal. Chem.* **56** (1984) 1953.
23. J. F. Pankow, M. P. Ligocki, M. E. Rosen, L. M. Isabelle, and K. M. Hart, *Anal. Chem.* **60** (1988) 40.
24. J. F. Pankow, L. M. Isabelle, and T. T. Kristensen, *J. Chromatogr.* **245** (1982) 31.
25. J. F. Pankow and L. M. Isabelle, *J. Chromatrogr.* **237** (1982) 25.
26. C. Leuenberger and J. F. Pankow, *Anal. Chem.* **56** (1984) 2518.
27. V. Leoni, G. Puccetti, R. J. Colombo, and A. M. D'Ovidio, *J. Chromatogr.* **125** (1976) 399.
28. V. Leoni, G. Puccetti, and A. Grella, *J. Chromatogr.* **106** (1975) 119.
29. J. J. Sun and J. S. Fritz, *J. Chromatogr.* **522** (1990) 95–105.
30. G. Grienberger, Ph.D. dissertation, Johannes Kepler Univ., Linz and Univ. of Innsbruck, Austria, 1995, p. 107.
31. M. Dressler, *J. Chromatogr.* **165** (1979) 167.
32. Anonymous, *Standard Methods for the Examination of Water and Wastewater*, 13th ed., American Public Health Association, New York, 1971, p. 259.
33. P. Van Rossum and R. G. Webb, *J. Chromatogr.* **150** (1978) 381.
34. C. D. Chriswell, R. L. Ericson, G. A. Junk, K. W. Lee, J. S. Fritz, and H. J. Svec,
*J. Am. Waterworks Assoc.* (Dec. 1977) 669.
35. A. Tateda and J. S. Fritz, *J. Chromatogr.* **152** (1978) 329.
36. H. Cohn, C. Eon, and G. Guiochon, *J. Chromatogr.* **119** (1976) 41–54.
37. C. Borra, A. DiCorcia. M. Marchetti, and R. Samperi, *Anal. Chem.* **58** (1986) 2048.
38. A. Dicorcia and R. Samperi, *Anal. Chem.* **62** (1989) 1490.
39. A. DiCorcia and M. Marchetti, *Anal. Chem.* **63** (1991) 580.
40. J. H. Knox, B. Kaur, and G. R. Millward, *J. Chromatogr.* **352** (1986) 3–25.
41. V. Coquart and M. C. Hennion, *J. Chromatogr.* **600** (1992) 195–201.
42. M. C. Hennion and V. Coquart, *J. Chromatogr.* **642** (1993) 211–224.
43. E. M. Flanigen and R. W. Grose, U.S. Patent 4,061,724 (Dec. 6, 1977).
44. G. M. W. Schultz-Sibbel et al., *Talanta* **29** (1982) 67.
45. I. Ogawa and J. S. Fritz, *J. Chromatogr.* **329** (1985) 81.
46. C. S. Oulman and C. D. Chriswell, U.S. Patent 4,277,635 (July 7, 1981).
47. T. G. Andronikashvili et al., *Chromatographia* **38** (1994) 613.
48. P. J. Dumont, Ph.D. thesis, Iowa State Univ., Ames, IA, 1995.
49. Dianna Mayer, P. J. Dumont, and J. S. Fritz, *J. Chromatogr. A* **771** (1997) 45.
50. P. A. Mills, *J. Assoc. Off. Anal. Chemists* **42** (1959) 734; **44** (1961) 171.

CHAPTER 4

# PRACTICAL CONSIDERATIONS: EQUIPMENT AND TECHNIQUES

## 4.1 THE FOUR STEPS OF SOLID-PHASE EXTRACTION

The SPE process can be divided into four main steps: (1) conditioning, (2) adsorption, (3) washing, and (4) elution. These are illustrated in Figure 4.1 (1).

One additional step is, of course, necessary to complete the analytical process—that of analysis. Most frequently the analysis step is performed by injecting a portion of the eluate into a gas or liquid chromatograph for separation and measurement of the individual sample components.

### 4.1.1 Conditioning

Before adsorption of analytes by the stationary phase can begin, the sorbent bed must be prepared and made compatible with the liquid solution. For example, in an extraction of hydrophobic substances from an aqueous medium it is clear that there must be close contact between the apolar phase (such as C18 silica) and the polar solution. Without some pretreatment the polar liquid flows in small channels through the solid phase without making the necessary close contact. The necessary pretreatment involves the use of a mediating solvent that will promote better surface contact between the phases.

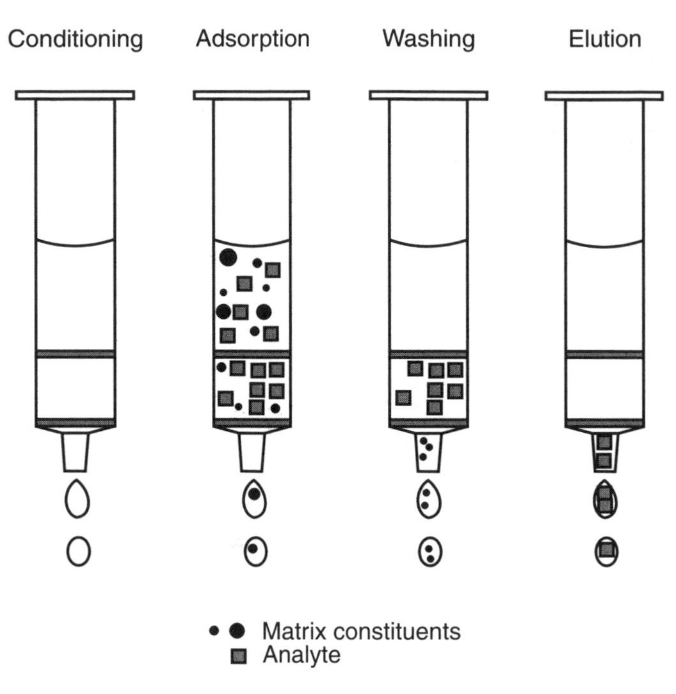


**FIGURE 4.1** The four steps in solid-phase extraction (1).

In the SPE of pesticides with C18 silica, it is customary to first activate the solid phase by addition of *n*-hexane to the column. In the dry form the C18 chains tend to be coiled up. Treatment with *n*-hexane causes these chains to uncoil, as shown in Figure 4.2 (1). When the stationary phase has become activated, it becomes almost transparent in appearance. After addition of methanol to displace the *n*-hexane and to fill up the pores, followed by a brief water wash, the desired conditioning has been achieved. There is good surface contact between the water and solid phases, and the C18 chains stick out like tentacles into the solution for optimum sorption of the hydrophobic analytes.

When a PS-DVB solid extractant is used, the conditioning step is simplified or can be eliminated entirely. The surface of PS-DVB particles is sufficiently hydrophobic to require a brief treatment of a mediating solvent such as methanol. With a surface modified resin such as acetyl PS-DVB recoveries of typical analytes are almost as good without any pretreatment as with a methanol pretreatment.

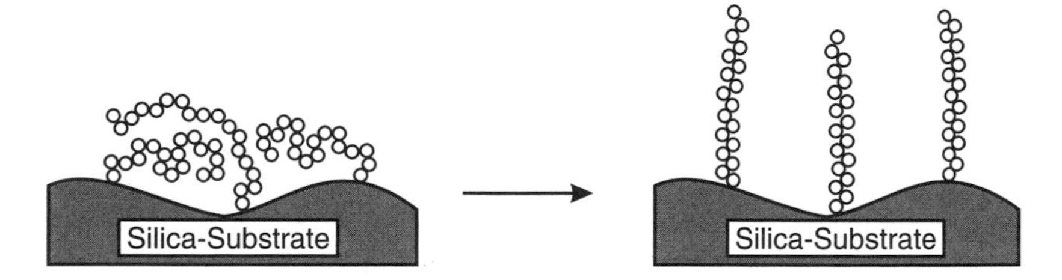


**FIGURE 4.2** Activation of a C18 phase with *n*-hexane.

### 4.1.2 Adsorption

The liquid sample to be extracted is passed through the packed column with the aid of suction (gentle vacuum), applied pressure, or a pump. Flow through the minicolumn should be at a reasonably constant rate. The flow rate used will depend on the column dimensions and on the particle size of the solid extractant. Very small particles (e.g., ~10 μm) are more efficient than columns packed with larger particles (e.g., ~50 μm) and thus permit a faster flow rate. A shorter sorbent bed can also be used with particles of smaller diameter. In fact, a column < 10 mm in length is usually appropriate for ~10 μm particles in order to avoid a high backpressure that would hinder liquid flow through the column. Columns or cartridges packed with 50–100 μm particles generally necessitate the use of a resin bed 50 mm or more in length and a slower flow rate to achieve efficient mass transfer.

How can one decide on an appropriate flow rate for a particular SPE column? One way is to use a preliminary test sample containing a highly colored solute. The tightness and length of the colored band on the SPE column will indicate whether the flow rate is suitable. A broad band can be tightened by using a slower flow rate. The colored band will also be narrower and more intense in color if sorbent of smaller particle size is used.

It should be noted that the sorbent bed should not be allowed to go dry at any point during the SPE process. The presence of air in the column prevents efficient interfacial contact between the liquid and solid phases.

### 4.1.3 Washing

A carefully chosen wash liquid affords the opportunity to remove coadsorbed matrix materials from the SPE column. (See the schematic drawing in step 3 of Figure 4.1.) What constitutes the right choice of a wash solution is problematic. For example, water ought to wash inorganic ions from a solid sorbent, but it may not be sufficient to remove other sample matrix components that are weakly adsorbed. In such cases, water containing 5–10 or 20% of an organic solvent, might be a suitable wash liquid. Of course, the solution used in the wash step must not contain a percentage of organic solvent high enough to partially elute the sample analytes.

### 4.1.4 Eiution

In the elution step the adsorbed analytes are removed from the solid extractant and are returned to a liquid phase that is suitable for analytical measurement. Most commonly, the eluting phase is an organic liquid, although it is often possible to thermally desorb analytes with the aid of a gas stream.

It must be kept in mind that the SPE column will usually be filled with water at the conclusion of the adsorption step. Thus, elution with an organic solvent will produce an effluent containing both water and the organic solvent. If the eluting solvent is not miscible with water, the effluent will contain two liquid phases. Even if the two liquids are compatible, the eluate will be more dilute (with regard to analytes) than might be wished.

For these reasons it is common practice to remove as much of the water as possible from the column just before the elution step. This can be accomplished by applying gentle vacuum for a few minutes or by passing compressed air or nitrogen through the column. Occasionally centrifugation is used to remove liquid from the column.

The eluting solvent should initially be added slowly and carefully to the now dry resin bed in order to avoid channeling. The eluting liquid should be chosen carefully. The most important thing is to select a liquid that will elute the analytes completely from the solid phase using as small a volume as possible of the eluent. In terms of capacity factor, this means that $k$ of the analytes should be as near as possible to zero.

There are several other important considerations. The eluting solvent must be compatible with the analytical measurement methods to be used. For example, when gas chromatography is to be used, the eluting solvent should have a fairly low boiling point so that the large solvent peak will not interfere with the sample peaks. The eluting solvent should be mostly free from impurities that might give disturbing chromatographic peaks. Finally, it should be low in cost and nontoxic. Proper disposal of toxic organic wastes is becoming a costly proposition.

The properties and relative elution strengths of organic solvents used in SPE are discussed in some detail later in this chapter.

## 4.2 APPARATUS

### 4.2.1 Cartridges and Packed Tubes

A variety of prepacked cartridges are available from supply houses for use in SPE. These cartridges are typically molded from solvent resistant polypropylene with very low levels of extractable impurities. The inlet end of the cartridge has a female luer taper for easy attachment to a large syringe (or other reservoir) containing the sample (see Fig. 4.3). The outlet has a standard male luer taper that can be attached to an outlet needle if desired. The packing is held in place by 20-μm polyethylene frits at each end of the cartridge.

Typical cartridges contain 300, 600, or 900 mg of packing bed. Available reversed-phase packings include C18, C8, C2, and phenyl bonded-phase silica particles. Normal-phase packings include diol, cyanopropyl, and aminopropyl silica as well as nonbonded silica and Florisil. As a very general guideline, cartridges can retain up to about 1 mg of sample components per 100 mg of packing. It is not advisable to use a cartridge with too large a packing bed, as this may result in incomplete elution of the desired compound. For compounds that are difficult to elute or that elute slowly, it is often advisable to perform the elution in the direction opposite that used for the original extraction. With reversed flow a compound that is retained near the top of the cartridge has a shorter path for elution from the cartridge.

SPE cartridges are popular, are easy to use, and work well for many purposes. However, their general design and packing efficiency are not always the best. Minicolumns are easy to pack and use, and they can be more efficient than cartridges.

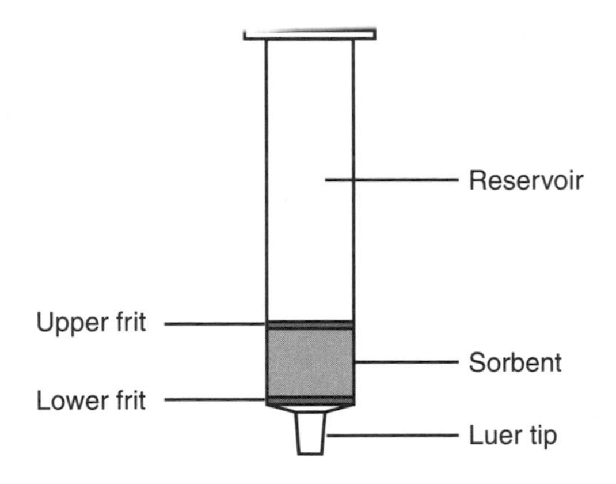


**FIGURE 4.3** Construction of an extraction column.

**TABLE 4.1 Resistance of Polypropylene to Various Solvents**

|  |  |
|---|---|
| Resistant | Methanol, ethanol, *n*-butanol, acetone, acetonitrile, DMF, hexane, cyclohexane, glacial acetic acid |
| | Hydrochloric acid (30%), phosphoric acid (80%), ammonium hydroxide (25%), sodium hydroxide (1 M), potassium hydroxide (1 M) |
| Somewhat resistant | Benzene, toluene, carbon tetrachloride, trichloroethylene, diethylether, tetrahydrofuran (THF) 1,4-dioxane, ethylacetate |
| Not resistant | Methylenechloride, chloroform, xylene, nitric acid (65%) |

Simple minicolumns can be easily packed with any desired sorbent using commercially available plastic housings. For example, a Varian Bond-Elut reservoir measures approximately 55 × 5.5 mm. It is flared out at the top for easy attachment to a syringe and has a male luer tip at the bottom. A circular polypropylene frit approximately 1 mm thick holds the resin bed at the bottom. After packing, a second frit can be placed on top of the resin bed to hold it in place and prevent entering liquid from disturbing the packing particles.

The stability of the column material to various organic eluting solvents is an important point to consider. Although polypropylene columns are prepared from a very resistant material, it is not acceptably stable in the presence of all types of solutions. In particular, contact with methylene chloride causes alkanes and plasticizers, such as 2,6-di-*tert*-butylcresol to be extracted from the column walls and from the polypropylene frits. These substances can lead to problems later in the analysis of overlapping- or interfering chromatographic peaks. Table 4.1 lists the resistance of polypropylene towards elution of impurities by various liquids.

A rather detailed study was reported concerning the impurities eluted from various frits and cartridges (see Tables 3.1 and 3.2). In cases where compatibility of polypropylene or other plastics might be a problem, glass columns can be used in conjunction with frits made of a more resistant material.

### 4.2.2 Resin-Loaded Membranes

A new generation of devices for SPE has recently emerged. These have the disk configuration of membrane filters but are loaded with sorptive resins or bonded-phase silica particles. These membranes include flat disks with high cross-sectional area that provide advantages not found in cartridges and packed mini columns. These disks have a low backpressure, which makes very high flow rates possible, and their wide bed and small thickness decrease the chance of plugging. New technologies for embedding the particles within the disk prevent channeling and improve mass transfer.

Chapter 6 is devoted entirely to disk technology in solid-phase extraction.

### 4.2.3 Filtration Apparatus

If the particle size of the packing material is not too small, it may be possible to pass the liquid sample through the SPE cartridge or column with simple gravity flow. However, it may be difficult to maintain a suitable flow rate under these very simple conditions. It can be pressed for more rapid flow.

Another simple way to overcome the resistance of the packed tube to liquid flow is to apply air or nitrogen pressure to the liquid sample in the reservoir. A simple needle valve is used to regulate the pressure applied and hence the rate of flow. An applied pressure of 1 or 2 bars is usually sufficient to achieve the desired flow rate.

Pressure can help to provide better contact between the liquid and solid phases and thereby improve the efficiency of SPE process. For example, activation of a C18 solid prior to SPE of pesticides normally requires around 30 ml of hexane. A much smaller volume of hexane will suffice if the stopcock is closed and a pressure of 1 bar is applied momentarily. The pressure applied forces hexane into the resin pores more efficiently (6).

Solid-phase extraction vacuum manifolds that permit simultaneous filtration of 12 or 24 samples are available. A diagram of a simple apparatus of this type is shown in Figure 4.4. The vacuum needed for filtration can be obtained by a simple water pump or by a vacuum pump with suitable controls. In one commercial apparatus precise flow control through each tube is obtained simply by rotating independent screw-type valves built into the cover (2). This ensures that the packing of some tubes will not go dry while others are still draining.

When a sample of a larger volume is to be used, refilling the rather small reservoirs can require close attention. This manipulation can be reduced simply by placing a suction tube into the bottom of a larger sample container, as shown in Figure 4.5 (3).

When only a few milliliters of sample is to be passed through a SPE tube, the necessary pressure can be achieved very simply by a use of a $\geq 25$ ml syringe. The sample solution is added directly to the column reservoir. The

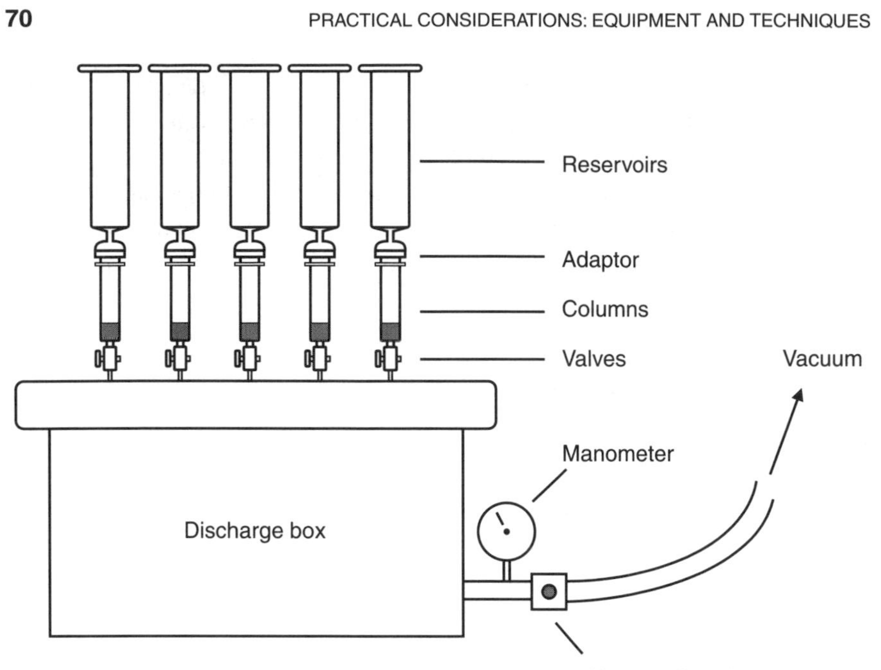


**FIGURE 4.4** SPE with vacuum technique.

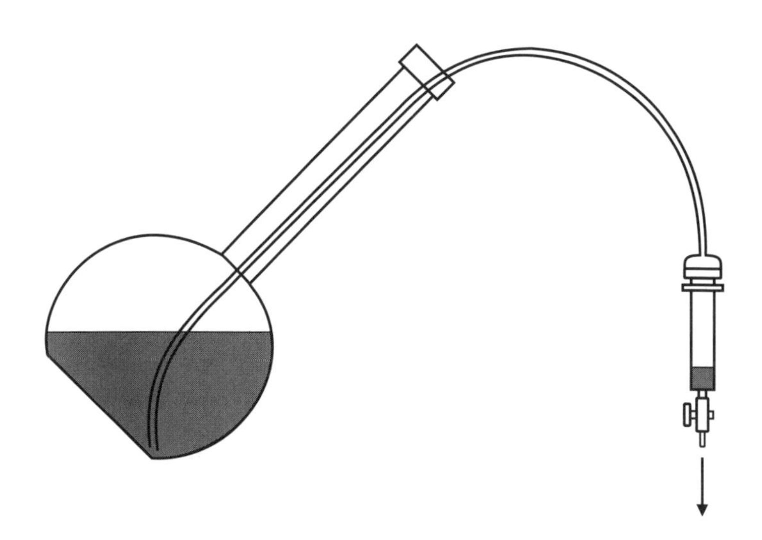


**FIGURE 4.5** Vacuum technique with large samples (3).

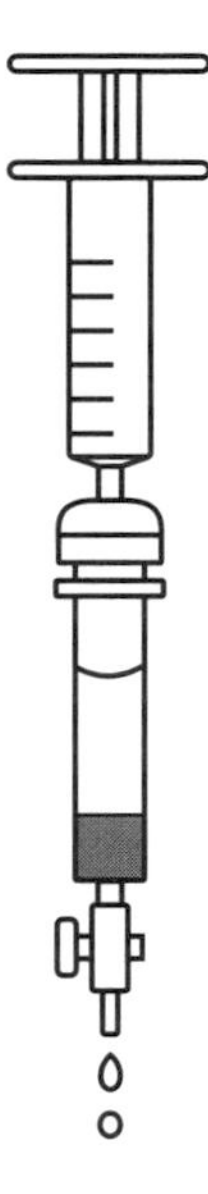

**FIGURE 4.6** Pressure technique with syringe.

syringe is filled with air and attached to the SPE column as shown in Figure 4.6 (4). The sample is then pushed through the SPE tube at the desired rate by applying pressure manually to the syringe piston.

A commercial single SPE tube processor uses a threaded piston (5). By rotating a knurled knob, pressure is applied smoothly for slow flow through the system. The plunger can be pressed for more rapid flow.

Another simple way to overcome the resistance of the packed tube to liquid flow is to apply air or nitrogen pressure to the liquid sample in the reservoir. A simple needle valve is used to regulate the pressure applied and hence the rate of flow. An applied pressure of 1 or 2 bars is usually sufficient to achieve the desired flow rate.

## 4.3 COMPLETENESS OF EXTRACTION

### 4.3.1 Breakthrough Volume

For efficient extraction to occur, distribution of an analyte between the sample liquid and the solid SPE phase must strongly favor the latter. This means that the distribution ratio, $D$, must be as large as possible [see Eq. (1.7)]. In column chromatography the distribution ratio, represented by the symbol $k$ (or sometimes $k'$), is called the *retention factor* or the *capacity factor*. In elution chromatography, $k$ is defined as the ratio of the *amount* (not the concentration) of analyte in the stationary phase to the amount in the mobile (liquid) phase. After introduction of the sample and pumping mobile phase continuously through the system, the analyte will emerge as a peak with the shape of a normal distribution curve. The retention volume, $V_R$, of this peak is given by the equation

$$V_R = V_0\,(1 + k) \tag{4.1}$$

where $V_0$ is the volume of liquid within the column and detector system.

In SPE the situation is a little different from that in elution chromatography. Now the sample itself is the mobile phase. When a sufficient volume of sample has passed through the SPE tube, the analyte begins to emerge from the tube in a concentration profile much like that of the chromatographic peak. However, in SPE the concentration of analyte does not reach a maximum and then decrease as it does in elution chromatography because more sample continues to flow into the system. The point at which the first analyte leaves the packed tube is called the *breakthrough volume*, $V_B$ (see Fig. 4.7) (6,7). The breakthrough volume is sometimes taken as the volume when the concentration of analyte leaving the tube constitutes 1% of its initial concentration in the sample.

The shape of a chromatographic peak depends on the number of theoretical plates ($N$) generated by the column:

$$N = 16\,\frac{V_R^2}{w} \tag{4.2}$$

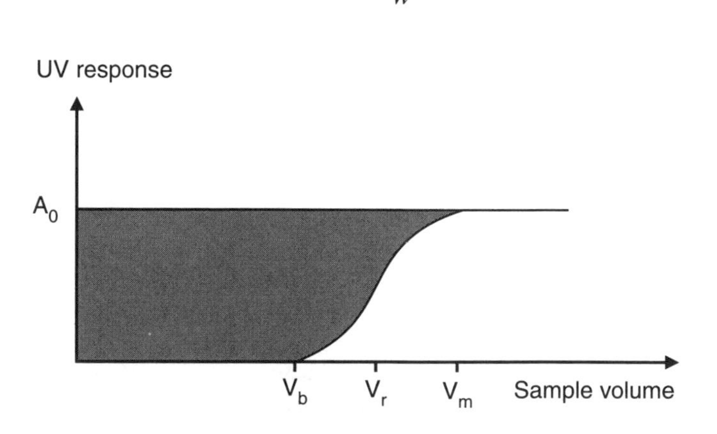


**FIGURE 4.7** Theoretical breakthrough curve obtained by percolation of a spiked sample (UV absorbance $A_0$) through a precolumn. See text for definintion of $V_b$, $V_r$, and $V_m$. (From Ref. 6 with permission.)

where $V_R$ is the retention volume of the analyte and $w$ is the width of the peak at its base ($w = 4\sigma$, where $\sigma$ is the standard deviation). For any given $V_R$ the peak will be sharper (smaller $w$) as $N$ increases. A sharp breakthrough profile is also advantageous for SPE because $V_B$ will occur later than when the breakthrough profile is more drawn out. Thus a larger number of theoretical plates in a SPE column is conducive to longer retention of analytes before breakthrough.

From this discussion it is apparent that two major conditions are needed for good SPE: a high retention factor ($k$) and a column with good partitioning efficiency. The latter can be represented by its number of theoretical plates ($N$). However, for a very short column such as an extraction tube, $N$ will not be very large even for an efficient system. Even so, the efficiency of most SPE tubes in terms of $N$ is not very good. The value of $N$ for a typical cartridge has been estimated to be 20 plates (7). There are several reasons for this. The particle size of a typical commercial SPE device is at least 40–50 μm, compared to particles of 3 or 5 μm in an HPLC column. Partition equilibrium of analytes between the liquid and solid phases is much slower with the larger particles, and $N$ is therefore smaller than for a comparable column packed with the smaller particles. The particle size range in SPE is much broader than in HPLC, leading to additional broadening of solute zones. Finally, SPE tubes are not as carefully packed as columns intended for chromatographic use. In short, SPE devices are just good enough to do their intended job.

Equation (4.1) indicates that the retention volume of an analyte depends strongly on its retention factor, $k$. For a substance to be extracted strongly and have a large breakthrough volume ($V_B$), it must have a large $k$. Values of $k$ vary tremendously with the chemical structure of the analyte as well as with the type of packing used in the SPE tube. Some typical values of $k$ for an assortment of chemical compounds on a bonded-phase silica column are listed in Table 4.2. These values for $k_w$ (retention factor in water) were selected from a paper by Braumann (8). The $k_w$ values listed are only approximate because they had to be obtained by linear extrapolation from methanol–water solutions, where the $k$ values were low enough to be measured experimentally, to pure water. Polar organic molecules have rather low $k_w$ values. As a result, their extraction by SPE tubes might be incomplete. At the other extreme, large fused-ring compounds such as anthracene and pyrene have very high $k_w$ values. These compounds will be extracted very strongly but complete elution could pose a problem unless an eluting solvent can be found where the $k$ value would be low enough to favor fast elution.

**TABLE 4.2. Retention Factors in Water ($k_w$) for Various Organic Compounds Using a Bonded-Phase Silica Column**

| Compound | log $k_w$ | $k_w$ | Compound | log $k_w$ | $k_w$ |
|---|---|---|---|---|---|
| Anilines | | | Halogenated benzenes | | |
| Aniline | 1.1 | 13 | Fluoro | 2.3 | 200 |
| 4-Nitro | 1.4 | 25 | Chloro | 2.8 | 630 |
| 4-Chloro | 1.8 | 63 | Dichloro | 3.5 | 3200 |
| | | | Trichloro | 4.4 | 25,000 |
| Phenols | | | Alkyl benzenes | | |
| Phenol | 1.3 | 20 | Benzene | 2.1 | 125 |
| Methyl | 1.8 | 63 | Methyl | 2.7 | 500 |
| Chloro | 2.3 | 200 | Dimethyl | 3.2 | 1,600 |
| Dichloro | 3.0 | 1,000 | Ethyl | 3.4 | 2,500 |
| | | | *n*-Propyl | 4.0 | 10,000 |
| | | | *n*-Butyl | 4.6 | 40,000 |
| Polar benzenes | | | | | |
| Acetophenone | 1.8 | 63 | | | |
| Benzyl alcohol | 1.3 | 20 | | | |
| Benzaldehyde | 1.7 | 50 | | | |
| Benzonitrile | 1.8 | 63 | | | |
| Nitrobenzenes | | | Fused-ring aromatics | | |
| Nitrobenzene | 1.9 | 80 | Naphthalene | 3.3 | 2,000 |
| 2-Nitro | 1.9 | 80 | Anthracene | 4.6 | 40,000 |
| 4-Nitro | 1.7 | 50 | Pyrene | 5.0 | 100,000 |
| 4-Methyl | 2.4 | 250 | | | |

*Source:* Data were selected from T. Braumann (8).

The breakthrough volumes of various analytes can be calculated (at least as a rough estimate) from $k_w$ values and column parameters. Suppose that we have an SPE tube with an approximate inside diameter of 0.9 cm packed with extractant particles to a total bed volume of 0.65 $cm^3$. Assume the dead volume, $V_0$, to be 0.4 $cm^3$ (0.4 ml). The retention volume, $V_R$, is easily calculated by substituting different values of $k_w$ into Equation (4.1). The width of a chromatographic elution peak at its base ($w$ or $4\sigma$) is calculated from Equation (4.2) using $N = 20$ as the estimated number of plates in the SPE tube. However, breakthrough in SPE will occur at a point $\frac{1}{2}$W (or $2\sigma$) from the retention volume. Thus the breakthrough volume, $V_B = V_R - 2\sigma$.

**TABLE 4.3 Calculated Values for Retention Volume ($V_R$) and Concentration Factor (CF) as a Function of Retention Factor ($k_w$)**

| $k_w$ | $V_R$ | $2\sigma$ | $V_B$ | CF |
|---|---|---|---|---|
| 50 | 20 | 9 | 11 | 7 |
| 100 | 40 | 18 | 22 | 14 |
| 500 | 200 | 89 | 111 | 70 |
| 1,000 | 400 | 179 | 221 | 138 |
| 2,000 | 800 | 358 | 442 | 276 |
| 3,000 | 1,200 | 537 | 663 | 414 |
| 5,000 | 2,000 | 894 | 1106 | 691 |
| 10,000 | 4,000 | 1,790 | 2,210 | 1,380 |

Breakthrough volumes for these conditions are calculated as a function of $k_w$ in Table 4.3. Note that $V_B$ is essentially a linear function of $k_w$. Thus, the amount of sample that can be safely put through an SPE tube depends strongly on the $k_w$ value of the analytes.

### 4.3.2 Concentration Factor

A larger sample also means that a greater degree of concentration is possible. We can define the concentration factor (CF) as the ratio of the volume of sample ($V_S$) to the volume of organic solvent used in the final eluting step ($V_E$):

$$\mathrm{CF} = \frac{V_S}{V_E} \tag{4.3}$$

The maximum $V_S$ that can be used is, of course, equal to $V_B$.

The concentration factor depends on the efficiency of elution as well as on the efficiency of extraction. A good eluting solvent might have a retention factor ($k$) < 1.0. For $k = 1.0$, the value of $V_B$ for the column described is

$$V_R = V_0\,(1 + \mathrm{k}) = (0.4)(2) = 0.8 \text{ ml}$$

However, twice this amount will be needed to elute the entire peak, giving an elution volume ($V_E$) of 1.6 ml. Concentration factors are also calculated in Table 4.3 using this value for $V_E$.

It should be remembered that one purpose of SPE is to isolate analytes from samples that contain salts or other substances that would interfere with subsequent measurement of the analytes by gas chromatography, or by other means. A large concentration factor is not always needed for this purpose. But another major use of SPE is to concentrate analytes from samples where their initial concentration is too low for adequate measurement. In this case a large concentration factor may be needed.

A good way to increase the concentration factor is to use sorbent particles that will have higher $k_w$ values than bonded-phase silicas. Figure 4.8 depicts plots of $k$ for benzene as a function of methanol concentration in aqueous methanol solution (6,7). The extrapolated $k_w$ values are higher for the polymeric PRP-1 than for the bonded-phase silica column packing. These plots also serve as a reminder that $k$, and hence $V_B$, will be lower in an aqueous sample that contains some methanol.

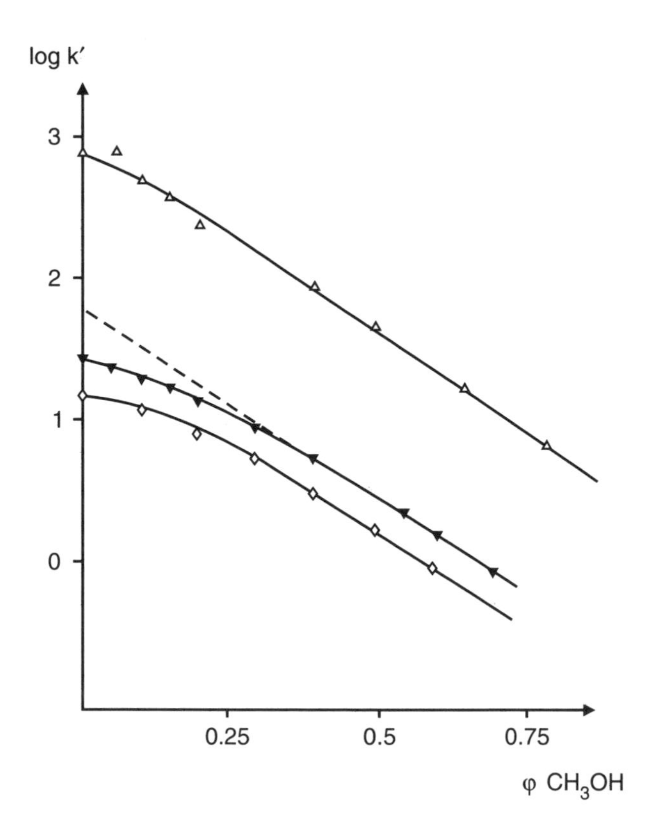


**FIGURE 4.8** Variation of the capacity factor of benzene with the volume fraction of methanol contained in the water–methanol mobile phase (▼ measured on a 10 × 0.46-cm column laboratory-packed with RP-8 silica; △ measured on a 5 × 0.46-cm column laboratory-packed with PRP-1 copolymer; ◇ measured on a 10 × 0.46-cm column packed with Hypercarb). (From Ref. 6 with permission.)

Data given in Table 3.12 indicate that $k$ values for typical organic analytes are almost always higher for a polymeric extractant such as PRP-1 than for alkyl bonded-phase silica materials. Modern carbonaceous materials such as pyrolyzed graphitic carbon (PGV) are often presented as being more retentive than either Ps-DVB or BP silica extractants. The data in Table 3.12 show this to be true for polar organic analyte such as phenols, but less polar compounds are more weakly retained by PGC than by the other two solid extractants.

## 4.4 ELUTION

### 4.4.1 Choice of Eluting Solvent

A considerable number of single organic solvents and mixtures of two or more solvents have been used for the elution step in solid-phase extraction. Methanol, acetone, acetonitrile, methylene chloride, and ethyl acetate have been used extensively for this purpose. Selection of an eluting solvent has often become a little like cooks exchanging their favorite recipes. Someone has found that a certain solvent works well for a given application, so others try it out for their own samples. For example, one desorption procedure advises the use of just enough acetone to cover the sorbent bed, followed by methylene chloride. This combination of solvents is easy to blow down (remove from the sorbent bed by air pressure) and, since the acetone proportion is <10%, it is easy to dry. These are both important practical considerations.

A more practical scientific approach would be to measure the capacity factors ($k'$) of various analytes on a small resin column using a pure organic solvent as the mobile phase. The capacity factor of a chemical should be as small as possible in order to obtain complete elution by a very small volume of the organic mobile phase. It might be assumed that most analytes would be so strongly solvated by a pure organic solvent (instead of the usual organic–water mobile phase used in HPLC) that the value of $k$ would approach zero. Unfortunately, this is not the case.

Table 4.4 compares the $k$ values of various compounds eluted from a tube containing PS-DVB particles with each of three pure organic solvents (9). The $k$ values for elution of these compounds from bonded-phase silica particles would be somewhat lower than from PS-DVB particles. The results in Table 4.4 show that acetonitrile is in general more effective than either methanol or ethanol for eluting most organic analytes. From Equation (4.1),

**TABLE 4.4 Capacity Factors for Various Compounds in Organic Solvents on Polystyrene–Divinylbenzene**

| Compound | MeOH | EtOH | ACN |
|---|---|---|---|
| Bromobenzene | 2.57 | 1.85 | 0.77 |
| Chlorobenzene | 1.99 | 1.24 | 0.60 |
| Benzonitrile[a] | 1.16 | — | 0.23 |
| Nitrobenzene | 1.80 | 1.63 | 0.30 |
| Benzaldehyde | 1.60 | 1.35 | 0.37 |
| Benzoic acid | 0.22 | 0.62 | 0.01 |
| Aniline | 0.60 | 0.66 | 0.22 |
| Naphthalene | 4.65 | 3.60 | 1.02 |
| 1-Chloronaphthalene | 6.40 | 4.52 | 1.62 |
| 1-Naphthol | 1.09 | 0.71 | 0.47 |
| Quinoline | 1.71 | 1.42 | 1.19 |
| Phenethyl alcohol | 0.46 | 0.29 | 0.16 |
| 3-Phenyl-1-propanol | 0.54 | 0.31 | 0.21 |
| Cinnamaldehyde | 2.91 | 2.22 | 0.43 |
| Phenylacetaldehyde | 1.40 | 1.31 | 0.28 |
| Cinnamylalcohol | 0.62 | 0.40 | 0.24 |
| Benzylamine | 4.20 | 3.25 | 1.01 |
| Benzylchloride | 1.65 | * | 0.39 |
| Benzylbromide | 2.12 | 1.65 | 0.38 |
| Benzylcyanide | 1.13 | 1.17 | 0.14 |
| Benzylmercaplan | 2.65 | 2.08 | 0.48 |
| Benzylacetate | 1.69 | 1.50 | 0.23 |
| Benzylacetone | 1.59 | 1.35 | 0.23 |
| Phenyl acetic acid | 0.34 | * | 0.01 |
| Benzene | 1.42 | 1.13 | 0.40 |
| Toluene | 1.86 | 1.42 | 0.50 |
| Ethylbenzene | 2.15 | 1.45 | 0.57 |
| Propylbenzene | 2.57 | 1.49 | 0.66 |
| Bulybenzene | 3.23 | 1.69 | 0.80 |
| Phenol | 0.37 | 0.26 | 0.15 |
| *p*-Cresol | 0.46 | 0.31 | 0.18 |
| 4-Ethylphenol | 0.53 | 0.32 | 0.21 |
| 4-Propylphenol | 0.61 | 0.33 | 0.26 |
| 4-*n*-Butylphenol | 0.73 | 0.37 | 0.32 |
| 4-*n*-Amylphenol | 0.91 | 0.42 | 0.39 |
| 4-*n*-Heptylphenol | 1.44 | 0.53 | 0.57 |
| *p*-Chlorotoluene | 2.67 | 1.82 | 0.75 |
| 2-Nitrotoluene | 2.04 | 1.87 | 0.33 |

[a]EtOH data not collected for this solute/solvent system.

*Source:* From Reference 9 with permission.

the retention volume ($V_R$) for elution will be 2 $V_0$. The minimum volume for complete elution will be 2 $V_R$, or 4 $V_0$, when $k = 1$.

### 4.4.2 Chemical Structure of Analyte

Several workers have studied the effect of a chemical substituent on the retention of a given solute in chromatography with an organic mobile phase (10,11). The chemical substituent contribution is frequently given as $T_x$, which is defined as

$$T_x = \log k_{R-x} - \log k_{R-H}$$

where $x$ represents the chemical substituent. The $\tau$ values for several substituents added to a benzene ring are given in Table 4.5 (9). As might be expected, a polar substituent ($-CO_2H$, $-OH$, $NH_2$, etc.) substantially re-

**TABLE 4.5 Functional Group Contributions to $\tau$ Values for the Benzene Ring**

| | Contribution, $\tau_x$ | | |
|---|---|---|---|
| Functional Group | Methanol | Ethanol | Acetonitrile |
| Br | 0.26 | 0.21 | 0.28 |
| Cl | 0.15 | 0.04 | 0.17 |
| CN | –0.09 | —[a] | –0.25 |
| $NO_2$ | 0.10 | 0.16 | –0.14 |
| CHO | 0.05 | 0.08 | –0.03 |
| COOH | –0.81 | 0.26 | –1.69 |
| $NH_2$ | –0.37 | –0.23 | –0.26 |
| $CH_2NH_2$ | 0.47 | 0.46 | 0.40 |
| $CH_2Cl$ | 0.07 | —[a] | –0.01 |
| $CH_2Br$ | 0.17 | 0.17 | –0.03 |
| $CH_2CN$ | –0.10 | 0.01 | –0.46 |
| $CH_2SH$ | 0.27 | 0.27 | 0.08 |
| OH | –0.58 | –0.63 | –0.44 |
| $CH_2COCH$ | 0.05 | 0.08 | –0.25 |
| $CH_2COOCH_3$ | 0.08 | 0.12 | –0.24 |
| $CH_2COOH$ | –0.62 | —[a] | –1.69 |

[a]Data not available.

*Source:* From Reference 9 with permission.

duces the value of $k$ single bonds while halogens, alkyl groups, and aromatic rings have the opposite effect.

Many of the compounds listed in Table 4.4 have $k'$ values considerably higher than 1.0 when either methanol or ethanol is the eluting solvent. In such cases the compound would require a comparatively large volume of solvent for elution.

The capacity factors of several polynuclear aromatic (PAH) compounds were measured on a PS-DVB resin column using seven different eluting solvents. The results, shown in Table 4.6, show that methanol and ethanol are very poor eluting solvents for PAH compounds. Acetonitrile would also be unsatisfactory for elution of anthracene and chrysene from a SPE column. Several other solvents are seen to have much lower capacity factors for the PAH compounds.

These results show that tetrahydrofuran (THF) and methylene chloride ($CH_2Cl_2$) provide the lowest capacity factors. However, for practical SPE methylene chloride has the disadvantage of not being miscible with water and thereby possibly being less efficient for elution after having passed an aqueous sample through the column. Ethyl acetate also gives low capacity factors and has the property of being quite volatile when gas chromatography is used to separate and measure the eluted sample components.

The data in Table 4.6 show that methanol is a very poor solvent for eluting fused-ring aromatic hydrocarbons and that ethanol and acetonitrile are also rather weak eluants. A chromatogram showed that benzene and naphthalene are readily eluted by methanol but anthracene elutes much later in a very broad peak. Chrysene failed to elute in any reasonable time. These results suggest that PAH compounds with more than two rings can be isolated

**TABLE 4.6 Chromatographic Capacity Factor for Selected Solutes in Pure Solvents**

| | Solvent | | | | | | |
|---|---|---|---|---|---|---|---|
| Solute Type | MeOH | EtOH | ACN | THF | EGME | EtOAc | $CH_2CH_2$ |
| Benzene | 1.44 | 1.13 | 0.40 | —[a] | —[a] | —[a] | —[a] |
| Naphthalene | 4.65 | 3.60 | 1.02 | 0.53 | 1.78 | 0.90 | 0.59 |
| Anthracene | 30.07 | 20.65 | 5.61 | 0.52 | 2.65 | 1.08 | 0.58 |
| Chrysene | —[b] | —[b] | 10.85 | 0.49 | 3.12 | 1.20 | 0.78 |

[a]Very low values.

[b]Very high values.

*Source:* From Reference 9 with permission.

selectively by solid-phase extraction. Smaller, more polar compounds in methanol will pass through a short, polymeric SPE column while the larger PAH compounds will be strongly retained. These, however, are quickly eluted by THF, ethyl acetate (EtOAc), or methylene chloride.

A simple experiment was tried in which a mixture of 75 μg/ml toluene, 1 μg/ml anthracene, and 5 μg/ml chrysene in ethanol–water was forced through a small cartridge containing 5 μm PS-DVB resin using positive air pressure. The minicolumn was washed with 0.5 ml methanol to remove the toluene. Then 1 ml tetrahydrofuran was used to elute the PAH compounds. Analysis of this fraction by HPLC showed no toluene peak and high recoveries of the two PAH compounds.

## 4.5 SOLID SAMPLES AND SLURRIES

### 4.5.1 Pretreatment Methods

In the analysis of solid samples such as vegetable matter, food, and animal tissue for pesticides or other organic chemicals, a sample pretreatment/concentration step is obviously called for. Frequently, the sample is ground up in a blender with a liquid to produce a fine slurry. This slurry or the solid sample, itself, is then extracted with an organic solvent or an organic–aqueous solvent mixture. Acetone–water or acetonitrile–water, containing in each case 60–90% v/v organic solvent, is widely used. The purpose of this extraction is to remove the organic analyte from the solid sample, but certain portions of the sample matrix are likely to be coextracted. After this extraction, the solid will often settle out so that a measured portion of the clear liquid can be drawn off. Otherwise, a filtration becomes necessary.

The next step is to separate and concentrate the organic analytes from the liquid extract by SPE using an apolar solid phase. This requires dilution of the organic–aqueous liquid with a large amount of water; otherwise the sample analytes will not be retained by the SPE column. For pesticide analysis it is recommended that the liquid be diluted until it contains no more than 3–10% of the organic solvent. For example, if 100 ml of 80% acetone is used for extraction of the solid sample, the final volume after dilution with water will be around 2 liters.

It is entirely feasible to subject a sample of ~2 liters to SPE, provided a sufficiently large reservoir is used. The time needed to pass this large sample volume through the SPE apparatus will be longer than for smaller samples. The large dilution of the acetone or acetonitrile with water reduces the solvating strength of the liquid for the rather hydrophobic analytes. This introduces the danger that some of the sample analytes will be adsorbed on the large surface areas of the reservoir and on the frits and connecting materials. Any adsorption on surfaces is likely to increase as the contact time becomes longer.

### 4.5.2 Fluid-Injection SPE

The problems associated with large dilutions of primarily organic solvent solutions with water are alleviated by use of a technique called fluid-injection SPE (FI-SPE). This uses an apparatus that permits "solid-phase extraction with continuous extract mixing" (7). In this method, dilution of the organic liquid with distilled water takes place in a continuous process just before contact with the SPE bed. This permits the analytes to remain in a favorable liquid medium as long as possible and thus minimize any losses due to surface adsorption.

A schematic diagram of the apparatus used in FI-SPE is shown in Figure 4.9 (12). Distilled water enters the mixing chamber by means of a variable-speed peristaltic pump. An aliquot (e.g., 25 ml) of the organic sample extract is placed in a polyethylene syringe and forced into the mixing chamber at a much slower rate than the distilled water by means of an electric motor fitted with a worm gear. The latter pushes the plunger of the syringe at a slow but constant rate, thus forcing the organic liquid into the mixing chamber. (A variable-speed infusion pump, which is commercially available, can be used.) Typically, a 20-fold dilution of the sample extract with water is achieved.

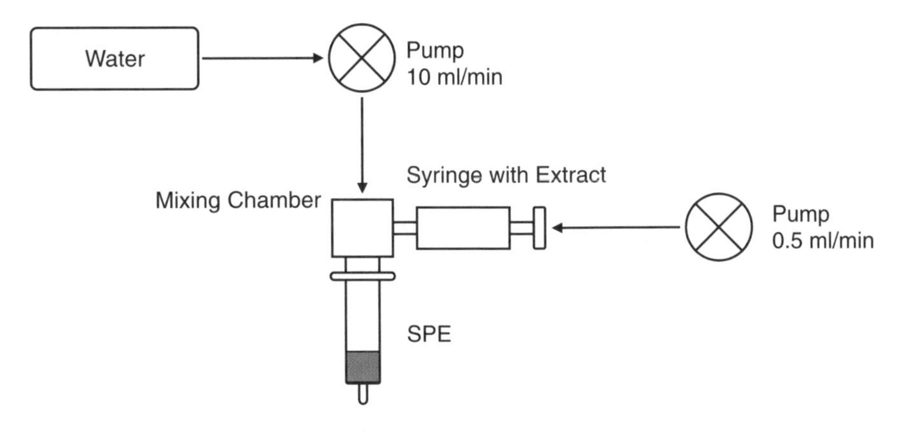


**FIGURE 4.9** Schematic drawing of FI-SPE (12).

The FI-SPE apparatus provides a very convenient, semiautomatic way to combine the dilution and SPE steps in a chemical analysis. Electronic control gives higher reproducibility of the flow rate and better reproducibility of the analytical results. A detailed description of the apparatus has been published by Grienberger (12).

## 4.6 AUTOMATION

### 4.6.1 General Approaches

In laboratories that have a large number of samples to analyze, automation has become essential. Various degrees of automation can now be applied to solid-phase extraction as well as to other analytical procedures. Automated SPE results in better precision. Operator errors are reduced by eliminating repetitive motions that cause fatigue and some lack of attention. Automated SPE also protects operators from the noise and crowding in production facilities and from hazards that may be present in other environments.

The following approaches have been used to make SPE faster and more convenient:

1. *Semiautomatic SPE.* Extraction plates are available with wells that hold 96 collection tubes, usually 12 rows of 8 tubes each. An array of 8 automatic pipettes applies samples to the system one row at a time.
2. *Workstations.* These instruments employ a certain amount of robotics. The instruments perform programmed operations such as load, rinse, and elute. A book by Thurman and Mills (13) contains an excellent review of the commercial work stations for SPE available at the time of publication (1998).
3. *On-line Systems.* This general method is illustrated by the system described in Section 4.6.3, in which samples are pumped through an SPE bed and the extracted substances are then eluted into a gas chromatograph that is coupled on-line with the SPE device.

### 4.6.2 Extraction Plate Systems

The Empore extraction plate systems (3M Co., St. Paul, MN, USA) is a good example of a modern system designed to speed up SPE. It is designed for parallel processing of 8, 12, or 96 samples at a time using a multichannel pipettor. The time needed to process 96 samples is reduced to <1 hour instead of an entire workday needed for SPE with 96 individual SPE tubes or cartridges.

A photo of the extraction plate system is shown in Figure 4.10. The top view shows the assembly with an extraction tube fitted into each of the 96 wells. The extraction tubes each contain an extraction disk (see Chapter 6) rather than a bed of loose extractant particles. A graded-density prefilter is placed on top of the extraction disk to remove any particulate matter and provide for faster, more consistent flow rates. A collar insert fits into the receiver tubes and helps prevent cross contamination.

The vacuum manifold cover is just below the assembly in Figure 4.10, and below that is a photo of standard 96-well collection vessels. After the sample has passed through the extraction tubes, the extracted sample components are eluted into the collection tubes. A shim (second to bottom in the figure) is inserted to provide proper spacing between the receiver and Empore disk plate. The entire system is connected to a vacuum source to provide for even liquid flow.

**FIGURE 4.10** Extractin plate system. (Courtesy of 3M Co.)

### 4.6.3 SPE Coupled On Line to a Gas Chromatograph

SPE is easier to use than liquid extraction and requires much less organic solvent (<1–5 ml) for desorption. However, often only 0.1–1% of the final extract is injected into a gas chromatograph. Thus, the aqueous sample volume has to be relatively large. Solid-phase microscale extraction (Section 8.1) avoids this limitation but the extraction step often is slow and incomplete.

The Brinkman research group in Amsterdam has been active in coupling SPE on line with the chromatographic system used to measure the sample components (14–18). This arrangement reduces significantly the amount of time needed for sample preparation. Since the total desorption volume is now introduced into the GC, the volume of aqueous sample can be made 10–100-fold smaller. Transfer of the desorption solvent into the gas chromatograph preferably is carried out with partial solvent evaporation. The interface of the SPE with the GC consists of a transfer capillary introduced into a retention gap via an on-column injector (19).

The scheme of the apparatus used is given in Figure 4.11 (16). The heart of the apparatus is a 10 mm × 10-mm-ID LC precolumn, packed with 10 μm PS-DVB resin (Polymer Labs PLRP-S). This acts as the SPE column.

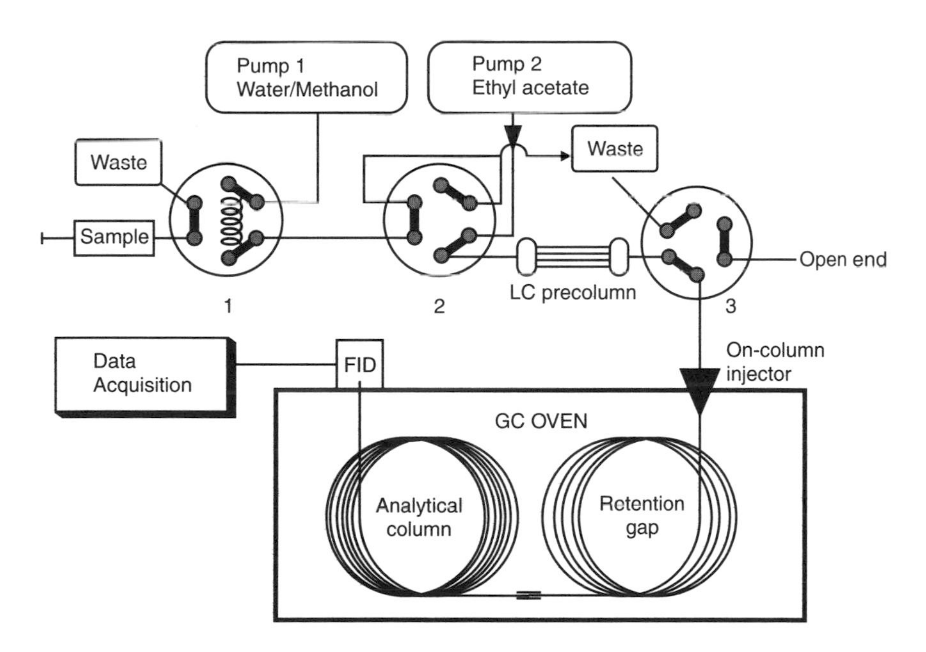


**FIGURE 4.11** Scheme of the system for SPE and on-line GC analysis. (From Ref. 16 with permission).

The system has a 0.92 ml loop for transfer of the aqueous sample to the SPE column. Valve 2 allows switching from aqueous preconcentration solvent to the desorption solvent, ethyl acetate. The desorption solvent is delivered by a pulse-free syringe pump. The SPE column is placed between valves 2 and 3, both of which are pneumatically actuated. Valve 3 transfers the desorption solvent either to waste or to a gas chromatograph via a 20 × 20 cm-ID fused-silica capillary. The latter penetrates through the on-column injector into the retention gap.

Sequential operations are controlled automatically with timing:

1. Methanol is added to the aqueous sample solution to make it 5% v/v in methanol. This is done to avoid sorption of analytes onto parts of the three-valve system. Analyte standards are prepared by adding measured amounts (spikes) of the analytes in methanol solution to aqueous samples.
2. The SPE column is conditioned by passing 2.25 ml of 95–5 water–methanol.
3. The sample is transported to the SPE column with water–methanol; the total volume of sample and additional solvent is 2.25 ml.
4. Valve 2 is actuated, and desorption solvent flow into the SPE column is started at a flow rate of 25 μl/min.
5. After 100 s, valve 3 is switched to start transfer to the GC system. The reason for this delay is to allow water to be purged from the SPE column before the nonaqueous desorption solvent is fed into the GC system. When 75 μl has been introduced (3.0 min), valve 3 is switched again and the transfer line is emptied by the purge flow.

Ethyl acetate is a good solvent for desorption of relatively polar compounds from C18 bonded-phase silica cartridges (20). Water is soluble in ethyl acetate to about 3%, at 20°C. Ethyl acetate saturated with water is suitable for elution of analytes from PS-DVB resins and for transfer to a GC retention gap (16).

When the entire volume of desorption solvent is transferred to a gas chromatograph for analysis, a much smaller aqueous sample can be used. In the SPE-GC analysis of water samples using a flame-ionization detector (FID), Brinkman et al. found that only 1 ml of water sample was needed to obtain detection limits the order of 0.1 ppb (0.1 μg/liter) for various analytes (16). With a nitrogen-phosphorous detector and a 2.5-ml water sample the

**TABLE 4.7 Average Recovery (Rec) and Mean Standard Deviation (MSD) for Seven Test Compounds at Three Concentrations ($n = 6$)**

| Test Compound | Rec ± MSD, % 0.1 ng/ml | Rec ± MSD, % 1.0 ng/ml | Rec ± MSD, % 10 ng/ml |
|---|---|---|---|
| Benzonitrile | 104 ± 6 | 98 ± 1 | 89 ± 3 |
| Nitrobenzene | 67 ± 11 | 100 ± 1 | 94 ± 1 |
| *m*-Cresol | 80 ± 17 | 92 ± 2 | 88 ± 1 |
| 2-Methylnaphthalene | 99 ± 2 | 92 ± 2 | 94 ± 4 |
| Acenaphthene | 98 ± 11 | 120 ± 3 | 109 ± 3 |
| 2,3-Dinitrobenzene | 81 ± 7 | 116 ± 6 | 106 ± 1 |
| Tributylphosphate | 140 ± 17 | 135 ± 6 | 113 ± 2 |

*Source:* From reference (16).

detection limits for several organophosphorus compounds was found to be in the low ppt (ng/liter) level.

Seven potential pollutants in surface water were selected to test the recovery and reproducibility of the on-line system. These test compounds varied rather widely in polarity and volatility. It was observed that the peak areas of all seven compounds did not change when the flow rate was varied from 90 to 900 μl/min during the preconcentration procedure. Excellent recoveries and good reproducibilities were obtained at each of three concentration levels in the original sample, as shown in Table 4.7.

## REFERENCES


1. G. Grienberger, Ph.D. Dissertation, Johannes Kepler Univ., Linz, Austria, 1995, p. 33.
2. *Supelco Chromatography Products* (catalog), 1994, p. 352.
3. G. Grienberger, Ph.D. dissertation, Johannes Kepler Univ., Linz, Austria, 1995, p. 88.
4. Ibid, p. 90
5. *Supelco Chromatography Products* (catalog), 1994. p. 355.
6. M. C. Hennion and V. J. Coquart, *J. Chromatogr.* **642** (1993) 211.
7. M. C. Hennion and V. Pichon, *Envir. Sci. Technol.* 28 (1994) 576A.
8. T. Braumann, *J. Chromatogr.* **373** (1986) 191.
9. T. K. Chambers and J. S. Fritz, *J. Chromatogr. A* **728** (1996) 271.
10. C. M. Riley, E. Tomlinson, and T. M. Jeffries, *J. Chromatogr.* **185** (1997) 197.
11. R. M. Smith, *J. Chromatogr. A* **656** (1993) 381.

12. G. Grienberger, Ph.D. dissertation, Johannes Kepler Univ., Linz, Austria, 1995, p. 93.
13. E. M. Thurman and M. S. Mills, *Solid-Phase Extraction, Principles and Practice*, Wiley, New York, 1998, Chap. 10.
14. E. Noroozian, F. A. Maris, M. W. F. Nielen, R. W. Frei, G. J. de Jong, and U. A. Th. Brinkman, *HRC & CC* **10** (1987) 17.
15. J. J. Vreuls, W. J. G. M. Cuppen, E. Dolecka, F. A. Maris, G. J. de Jong, and U. A. Th. Brinkman, *HRC & CC* **12** (1989) 807.
16. J. J. Vreuls, W. J. G. M. Cuppen, G. J. de Jong, and U. A. Th. Brinkman, *HRC & CC* **13** (1990) 157.
17. J. J. Vreuls, W. J. G. M. Cuppen, G. J. de Jong, and U. A. Th. Brinkman, *HRC & CC* **13** (1992) 1701.
18. U. A. Th. Brinkman, *Environ. Sci. Technol.* **29**(2) (1995) 79A.
19. F. Munari, A. Trisciani, G. Mapelli, S. Trestianu, K. Grob, and J. M. Colin, *HRC & CC* **8** (1985) 601.
20. J. M. Vinuesa, J. C. M. Cortés, C. I. Cañas, and G. F. Pérez, *J. Chromatogr.* **472** (1989) 365.

# CHAPTER 5

# ION-EXCHANGE SORBENTS

## 5.1 INTRODUCTION

Solid particles with cation-exchange groups or anion-exchange groups have found considerable use in SPE. For example, a strong-acid cation exchanger contains $-SO_3^-H^+$groups at the end of an organic chain on a bonded-phase silica or attached to the benzene ring of a polymeric resin. The $-SO_3^-$ part is attached by a chemical bond, but the $H^+$ counterion is held more weakly by electrostatic attraction. Another cation can displace the $H^+$and set up an ion-exchange equilibrium:

$$\text{Solid} -SO_3^-H^+ + Na^+ \rightleftharpoons \text{Solid} -SO_3^-Na^+ + H^+ \tag{5.1}$$

The equilibrium constant, $K$, is expressed by

$$K = \frac{\text{solid} - SO_3^-Na^+ \cdot H^+}{Na^+ \cdot \text{solid} -SO_3^-H^+} \tag{5.2}$$

Values of $K$ vary considerably but in general increase with the charge and bulkiness of the exchanging ion. Cation-exchange resins of this type can retain organic or inorganic cations by an ion-exchange mechanism, or they may react with a neutral base to form a cation. Thus, a neutral organic base (B), such as an amine, is converted by acid to a protonated cation, $BH^+$:

$$\text{Solid}\ -SO_3^-H^+ + B \rightleftharpoons \text{solid}\ -SO_3^-BH^+ \tag{5.3}$$

Another useful application of ion-exchange solid particles is the neutralization of acidic or basic ions without adding an ion of opposite charge to the sample. As an example, a strong-acid cation exchanger may be used to neutralize a high concentration of a basic anion while allowing lower concentrations of other sample anions to pass through the ion-exchange tube:

$$\text{Solid}\ -SO_3^-H^+ + Na^+OH^+ \rightleftharpoons \text{solid}\ -SO_3^-Na^+ + H_2O \tag{5.4}$$

$$\text{Solid}\ -SO_3^-H^+ + Na^+Cl^- + Na^+NO_3^- \rightleftharpoons \text{solid}\ -SO_3^-Na^+ + H^+Cl^- + H^+NO_3^- \tag{5.5}$$

### 5.1.1 Ion-exchange materials

Ion-exchange materials that are available commercially for SPE are listed in Table 5.1. In addition to the strong-acid cation exchangers, weak-acid cation exchangers are available with a carboxymethyl group ($-CH_2CO_2H$) bonded to the silica. This material acts as a cation exchanger only in solutions where the pH is sufficiently high (>5) for the carboxyl group to become ionized and therefore able to attract an analyte cation as the counterion.

Two major types of silica-based anion exchangers are available. Aminopropyl silica ($-CH_2CH_2CH_2NH_2$) becomes an anion exchanger where the sample pH is sufficiently low to protonate the amine group and thereby attract a sample anion: ($-CH_2CH_2CH_2N^+H_3A^-$). A strong anion exchanger (SAX) bears a trimethylaminopropyl group [$-CH_2CH_2CH_2N^+(CH_3)_3Cl^-$]. The counterion ($Cl^-$) is readily exchanged for a sample anion ($A^-$) on the resin exchange sites.

Polystyrene and polyacrylate resins can also be converted to ion exchangers by appropriate chemical reactions. Sulfonated polystyrene is a strong cation exchanger. Anion-exchange resins are found in two major types: type 1 has a benzyltrimethylammonium group [$-CH_2N^+(CH_3)_3Cl^-$], and type 2, which has a benzyldimethylhydroxyethyl structure: [$-CH_2N^+(CH_3)_2(CH_2CH_2OH)Cl^-$].

Polymeric ion exchangers are manufactured in two polymerization types. Microporous ion-exchange resins swell readily in water to become gels unless they have a sufficient degree of crosslinking. But too much crosslink-

**TABLE 5.1 Types of Ion Exchangers for SPE**

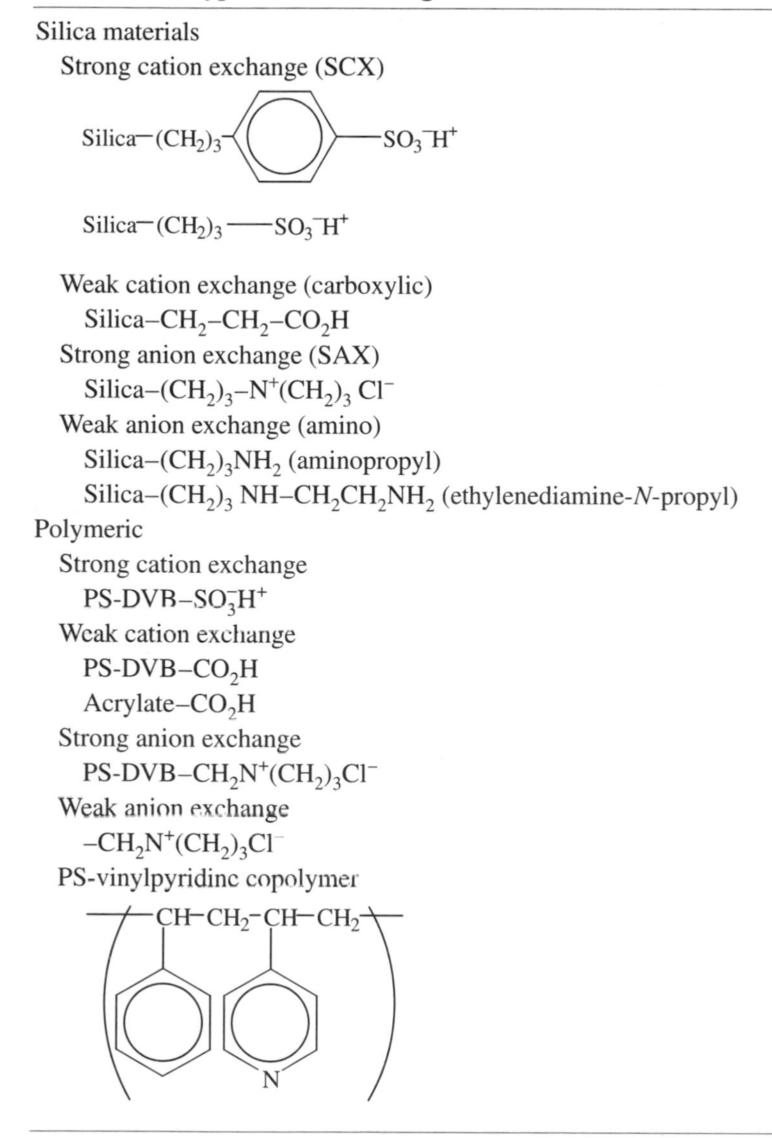

Silica materials
- Strong cation exchange (SCX)
  - Silica–$(CH_2)_3$–$C_6H_4$–$SO_3^-H^+$
  - Silica–$(CH_2)_3$–$SO_3^-H^+$
- Weak cation exchange (carboxylic)
  - Silica–$CH_2$–$CH_2$–$CO_2H$
- Strong anion exchange (SAX)
  - Silica–$(CH_2)_3$–$N^+(CH_2)_3$ $Cl^-$
- Weak anion exchange (amino)
  - Silica–$(CH_2)_3NH_2$ (aminopropyl)
  - Silica–$(CH_2)_3$ NH–$CH_2CH_2NH_2$ (ethylenediamine-*N*-propyl)

Polymeric
- Strong cation exchange
  - PS-DVB–$SO_3^-H^+$
- Weak cation exchange
  - PS-DVB–$CO_2H$
  - Acrylate–$CO_2H$
- Strong anion exchange
  - PS-DVB–$CH_2N^+(CH_2)_3Cl^-$
- Weak anion exchange
  - –$CH_2N^+(CH_2)_3Cl^-$
- PS-vinylpyridine copolymer
  - –[CH–$CH_2$–CH–$CH_2$]– (phenyl and pyridyl (N) substituents)

ing inhibits transport of ions within the resin bead. The gel structure tends to collapse when an organic solvent comes into contact with the resin. Macroporous resins (sometimes called *macroreticular resins*) are porous materials with a rigid polymeric structure. They usually have a high degree of crosslinking. However, ions to be exchanged can move readily through the resin beads by virtue of the many pores and channels within the resin.

These are effective ion exchangers in most organic solvents as well as in water. Macroporous resins are usually the best for SPE use.

## 5.2 ION-EXCHANGE SPE PROCESSES

Ion-exchange SPE involves two basic steps:

1. Uptake of sample analytes as ions
2. Elution of the analytes

### 5.2.1 Sample Ion Uptake

In the first step the liquid sample is added to a tube filled with appropriate ion-exchange particles. Sometimes the sample must be made either acidic or basic so that the analytes will be in their ionic forms. The sample ions are retained by the ion-exchange particles; they are held in place by electrostatic attraction with the oppositely charged exchange groups.

After the sample has passed through the SPE tube a small volume of wash solution is added to rinse nonionic sample materials from the tube. Water alone may be used as the sample and wash solvent for inorganic ions and small organic ions. However, it is often necessary to use a mixed aqueous–organic solvent in both the sample and the wash to prevent adsorption of neutral organic sample material. Ion-exchange uptake of sample ions is feasible in mixed solvents or even in pure organic solvents such as methanol or acetonitrile.

Some analytes are always present as ions. These include the following:

| | |
|---|---|
| Metal cations: | $Na^+$, $Ca^{2+}$, $Zn^{2+}$, etc. |
| Inorganic anions: | $Cl^-$, $NO_3^-$, $SO_4^{2-}$, etc. |
| Organic cations: | $R_4N^+$ |
| Organic anions: | $C_6H_5SO_3^-$, etc. |

These sample ions are extracted from the sample by simple ion exchange. For example, an organic sulfonic acid undergoes the following equilibrium with a solid anion exchanger with quaternary ammonium groups:

$$C_6H_5SO_3^- + \text{solid} -N^+Cl^- \rightleftharpoons \text{solid} -N^+C_6H_5SO_3^- + Cl^- \qquad (5.6)$$

Uptake of the sulfonate anion should be essentially quantitative for two reasons: (1) an organic anion of this type is more strongly attracted to the positively charged exchange sites than inorganic ions such as chloride, and (2) multiple equilibrations in the short SPE column push the exchange equilibrium to completion.

### 5.2.2 Elution

In this example, the solution used for the elution step must contain a sufficiently high concentration of an ion such as chloride to shift the equilibrium in Equation (5.6) back to the left. This could require the order of a 0.1 M concentration of NaCl or HCl for adequate elution. The other option for SPE of an organic sulfonate would be to use a weak-base anion exchanger rather than the quaternary ammonium exchanger. A weak-base exchanger would lose its proton, and hence its anion exchange capability at a highly alkaline pH. Elution with an aqueous solution of sodium hydroxide, with possibly some added methanol, should elute the sample anion efficiently from a weak-base ion-exchange column.

Similar considerations apply to SPE with a solid cation exchanger. Sample substances that are ionic at any pH will require an eluting solution of relatively high cation concentration when a strong-acid cation exchanger is used. However, a weak-acid solid with carboxylic acid groups has an estimated $pK_a$ of 4.8. This means that sample cations will be taken up at pH values above ~6 where the solid is mostly in the negatively charged carboxylate form. Elution is facilitated by use of a solution of pH below ~4 where the solid is mostly in the molecular –COOH form.

Many organic and some inorganic substances are in their ionic form only in a certain pH range. Several examples are listed below:

| pH 3 | pH 11 |
|---|---|
| $H_2CO_3$ | $CO_3^{2-}$ |
| HF | $F^-$ |
| $RCO_2H$ | $RCO_2^-$ |
| $C_6H_5OH$ | $C_6H_5O$– |
| $RNH_3^+$ | $RNH_2$ |
| $PyrNH^+$ | PyrN (pyridine) |

### 5.2.3 Selectivity

More subtle changes in sample pH may be used to provide additional selectivity in the SPE process. Analytes may differ with regard to the pH at which they become ionic. Consider, for example, an aliphatic amine (B) with an ionization constant in water of $k_B = 10^{-3.4}$ ($4.0 \times 10^{-4}$). The acidic ionization constant of the protonated amine ($BH^+$) is $k_a = 10^{-10.6}$ ($2.5 \times 10^{-11}$):

$$k_a = \frac{[H^+][B]}{[BH^+]} \tag{5.7}$$

At pH 10.6 ($[H^+] = 10^{-10.6}$) the protonated and molecular forms of the amine will be present in a 1:1 ratio. At pH 9.6 $[BH^+]:[B] = 10$ and at pH 8.6, this ratio is 100.

Thus, at any pH below ~9 the amine will be in predominately ionic form, $BH^+$, and it should be taken up by a cation-exchange resin. However, the ionic form of the ion exchanger will affect the $B–BH^+$ equilibrium. If the resin column is in the "hydrogen" form (resin–$SO_3$–$H^+$), the molecular form of the amine will be converted to $BH^+$ through contact with $H^+$ on the resin. This effect can be mostly nullified by first passing a buffer of the desired pH through the ion-exchange column or more simply by converting the column to the $Na^+$ or $NH_4^+$ form. Even when this is done, the B-$BH^+$ equilibrium in a buffered solution will be shifted somewhat by uptake of $BH^+$ by the ion exchanger.

Consider next a pyridine analyte with a $k_b = 10^{-9.0}$ ($k_a$ of $BH^+ = 10^{-5.0}$). Equilibrium calculations show that the ratio $BH^+$:B is 1:1 at pH 5.0 and 10:1 at pH 4. Theoretically, this analyte should be well taken up by a cation exchanger from solutions with pH < 4. But at what pH will this pyridine compound be in predominately basic form (B) and therefore not retained by the cation exchanger? Equilibrium calculations show that the $BH^+$:B ratio is 0.1 at pH 6 and 0.01 at pH 7.

From these calculations it is predicted that when the sample solution is at approximately pH 7–9, the aliphatic amine will be taken up by the cation-exchange column while the pyridine analyte will be in the B form and therefore not taken up. With resins containing a high concentration of polar sulfonate groups, B *is* likely to pass quickly through the column. However, resins with only a low concentration of polar exchange groups will often retain neutral analytes by adsorption. In such cases analytes can be retained by two mechanisms: ion exchange and adsorption. This situation will be discussed in the following sections.

### 5.2.4 Examples

Following the ion-exchange retention of the selected sample ions, the elution strategy is usually to incorporate an acid or base in the eluting solution to convert the analytes to their molecular, nonionic form. For example, triazine herbicides (atrazine, propazine, etc.) are present as protonated cations ($pK_a = 2$) in acidic solution. They are readily taken up by a strong-acid cation exchanger. Subsequent elution is accomplished by a 0.1 M solution of potassium phosphate (a strong base) in a 50:50 mixture of acetonitrile and water (1). The base converts the herbicides back to their molecular forms and the acetonitrile solvates the herbicides to prevent physical adsorption by the solid exchanger.

Solid-phase extraction of phenols by a strong anion exchanger is another example of the principles commonly used. Although the acidic strength of phenols varies over a broad range, depending on substituents, most phenols have $pK_a$ values of ~8–10. A pH > 10 must be used to convert such phenols to the anionic form needed for ion exchange. This may be done by adjusting the sample pH with sodium hydroxide, or the sample at a lower pH may be added directly to the anion-exchange column in the hydroxide form (the strong base of the ion exchanger neutralizes any anionic forms):

$$C_6H_5OH + \text{solid} -N^+OH^- \rightleftharpoons \text{solid} -N^+C_6H_5O^- + H_2O \qquad (5.8)$$

A dilute solution of HCl in methanol works well for the elution of phenols from the ion-exchange column. The acid converts the phenols back to their molecular form; the methanol causes the molecular phenols to be rapidly eluted. It must be noted that enough of the acidic methanol solution must be used to neutralize all of the hydroxide ions on the ion exchanger:

$$H^+Cl^- + \text{solid} -N^+OH^- \rightleftharpoons \text{solid} -N^+Cl^- + H_2O \qquad (5.9)$$

If insufficient HCl is used, the remaining $-N^+OH^-$ groups will retain phenolate anions farther down the column. To minimize the volume of eluting solution required, the exchange capacity of the SPE tube should be only enough to assure complete uptake of the phenols with a reasonable margin of safety. In addition, the acid concentration of the eluting solution should be reasonably high (perhaps ~0.1 M) to keep the volume of eluting solution as low as possible.

After an acidic solution has been used to elute phenols from an anion-exchange column, the column must be returned to its original hydroxide form before the next SPE run. This may be done by passing a dilute solution of sodium hydroxide through the tube, followed by a brief water rinse.

SPE of a weak organic base (B) follows the following outline:

Uptake:

$$B + \text{solid} -SO_3^-H^+ \rightleftharpoons \text{solid} -SO_3^-BH^+ \qquad (5.10)$$

Elution:

$$NaOH + \text{solid} -SO_3^-BH^+ \rightleftharpoons \text{solid} -SO_3^-Na^+ + B + H_2O \qquad (5.11)$$

Regeneration:

$$HCl + \text{solid} -SO_3^-Na^+ \rightleftharpoons \text{solid} -SO_3^-H^+ + NaCl \qquad (5.12)$$

The elution step may be done with a volatile base such as $NH_3$ or $CH_3NH_2$ in methanol instead of the nonvolatile sodium hydroxide.

### 5.2.5 Bimodal Analyte Uptake

Some ion-exchange materials can be used for SPE in either of two modes. One is the ion-exchange mode in which the sample substances are retained as either cations or anions. The other possibility is for neutral analytes to be retained by physical adsorption similar to that encountered in the previous chapters. The latter mode requires ion exchangers of just the right properties. Cation exchangers with a high density of sulfonic acid groups or anion exchangers with a high concentration of quaternary ammonium groups have strong ion-exchange properties but relatively weak ability to adsorb molecular analytes. By using materials with a lower density of polar exchange groups, uptake of neutral molecules is enhanced and sufficient capacity for ion exchange is still possible. The preparation and applications of sulfonated resins for SPE by either adsorption or ion exchange are discussed in the next section.

## 5.3 RETENTION OF NEUTRAL ANALYTES BY SULFONATED RESINS

### 5.3.1 Introduction

It was noted in the previous section that organic bases are retained as the protonated cations on sulfonated polymeric resins. The protonated amine cations are strongly attracted to the negatively charged sulfonate groups of the sulfonated resins. In addition there is an attraction of the hydrophobic parts of the amine for the apolar polystyrene matrix of the resin (2,3). Fritz and Schmidt (4,5) showed that neutral analytes as well as cations are retained by sulfonated resins provided the degree of sulfonation is not too great.

A major problem with many of the particles used in SPE is the inability of aqueous sample solutions to adequately wet the surface, which is usually hydrophobic. Pretreatment of the resin column or cartridge with methanol is usually necessary to obtain better surface contact with the aqueous solution. In Chapter 3, it was shown that introduction of polar groups into a polystyrene-DVB resin greatly increases the wettability of the resins and gives more efficient extraction, particularly for polar organic compounds. Fritz and Sun (6) modified PS-DVB resins with alcohol, aceto, and cyano groups. Dumont and Fritz (7) found that sulfonation of the resin surface resulted in good interfacial contact between aqueous sample solutions and the sulfonated resins.

### 5.3.2 Sulfonation of Resins

The goal of resin sulfonation was to make the resin surface more hydrophilic while keeping the remaining parts of the resin largely unchanged for extracting organic solutes (7). A fairly rapid sulfonation with sulfuric acid was used because sulfonation of resins is known to proceed from the outside into the resin. In order to achieve more even wetting of the resin with viscous conc. sulfuric acid, the spherical resin beads were first slurried with a little glacial acetic acid.

Porous PS-DVB resin beads (average size 8 μm) were sulfonated under a variety of conditions to produce sulfonated resins ranging in capacity from 0.1 to 2.7 meq/g. The sulfonation conditions and capacities are given in Table 5.2.

### 5.3.3 Measurement of Capacity Factor

The efficiency of resins used for SPE is most commonly determined by measuring the percentage recovery of test solutes. However, this process depends on the efficiency of elution of the analytes from the SPE column as well as the efficiency of the initial extraction step. A better way to compare the behavior of different resins is to measure the capacity (retention) factor of the extraction step. However, it is often difficult to measure $k$ in pure water when $k$ is often very large. One approach is to measure the capacity factor in several different organic–aqueous solvent mixtures. Then,

**TABLE 5.2 Reaction Conditions for Resin Sulfonation**

| Capacity, meq/g | $H_2SO_4$, ml | Reaction time, min | Temperature |
|---|---|---|---|
| 0.0 | | | |
| 0.1 | 5 | 0.5 | Ice |
| 0.4 | 50 | 2 | Ice |
| 0.6 | 50 | 4 | Ice |
| 1.0 | 50 | 10 | RT[a] |
| 1.2 | 50 | 20 | RT[a] |
| 1.5 | 50 | 90 | RT[a] |
| 2.1 | 50 | 90 | 50 °C |
| 2.7 | 50 | 90 | 85 °C |

[a]Room temperature.

by extrapolating a plot of log $k$ versus $\phi$, the percentage of organic solvent in the mobile phase, to $\phi = 0$, one can estimate the capacity factor in pure water. Curve fitting techniques may also be used. Mills et al. equilibrated a spiked aqueous sample with resin for 24 h; the analyte concentration remaining in the aqueous phase was used to calculate the equilibrium constant (8).

Dumont and Fritz (7) determined the $k$ values of various analytes by elution from a very small resin column using pure water as the eluant. The method is quick and convenient and requires no extrapolation because no organic modifier is used in the eluant. The capacity factor is determined from the recorded elution curve using the well-known relationship $k = (t_R - t_0)/t_0$.

A very small column must be used for this method to be feasible because the high values of $k$ necessitate an elution of many column volumes. The values of $k$ are also very dependent on the measured $t_0$, although an error into $t_0$ will still give relative values of $k$ that can be compared for different resins. Even with a very small column, nonpolar compounds such as benzene would be retained for several hours and give very flat elution "peaks." For this reason more polar, water-soluble compounds were chosen as test materials: phenol, catechol, 2,3-butanedione, and ethylpyruvate. These compounds are polar enough to elute in a reasonable time and are easily detected by a UV–vis detector.

The $k$ values of two of the four test compounds are plotted against the sulfonic acid capacity of the resins in Figure 5.1. In each case $k'$ increases with increasing resin capacity, reaching a maximum at about 0.6 meq/g.

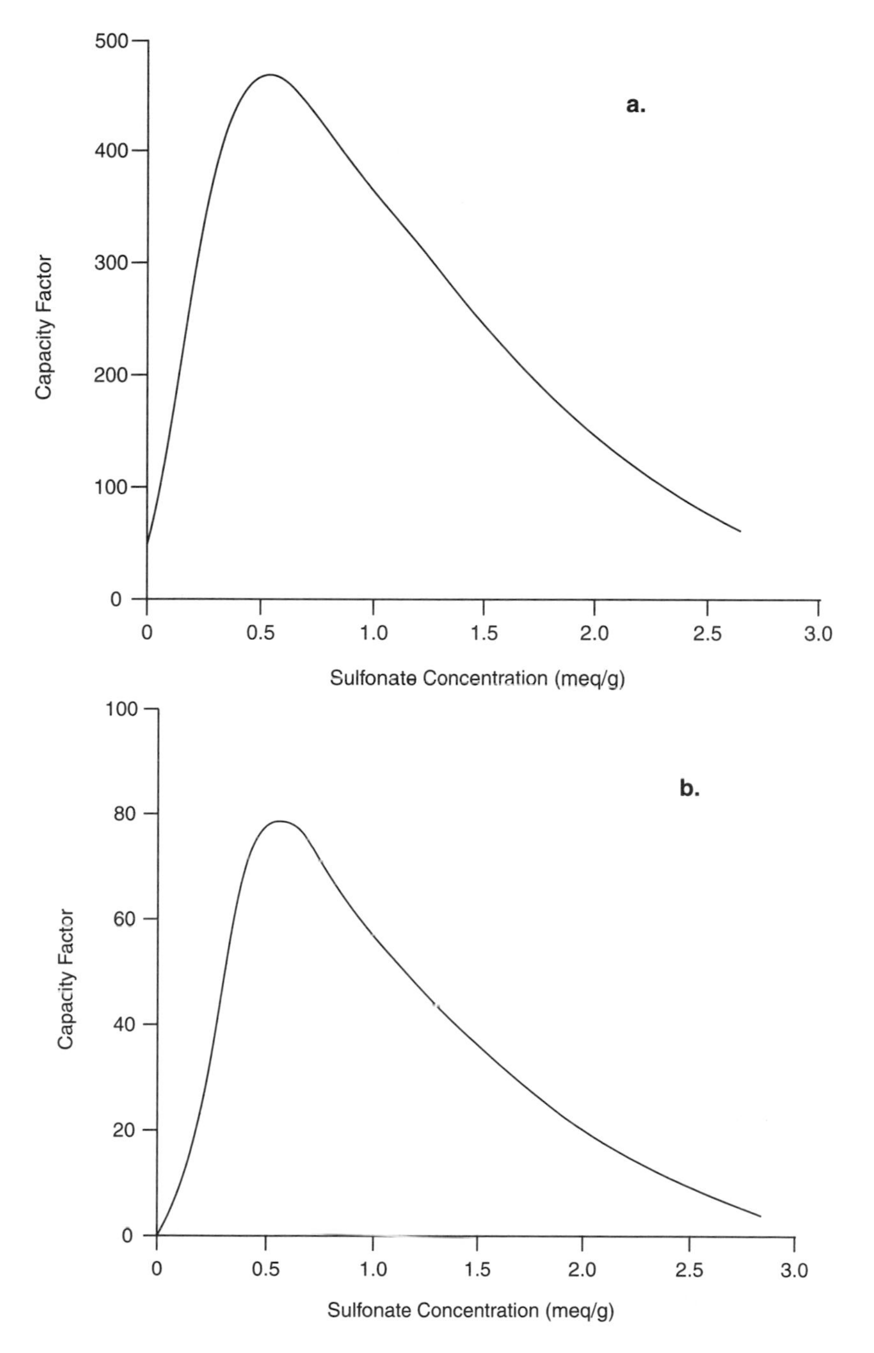


**FIGURE 5.1** Effect of resin sulfonate concentration on capacity factor: (*a*) phenol; (*b*) ethylpyruvate.

Further increases in sulfonic acid capacity are marked by a rapid decrease in $k$.

The increasing $k$ values of up to 0.6 meq/g can be attributed to the fact that a surface-sulfonated resin is more hydrophilic and therefore more easily wettable. The wettability of a resin may be quickly checked by adding a few milligrams of dry resin to water. Hydrophobic resins will remain on the surface of the water even if stirred. Hydrophilic resins will be dispersed throughout the solution because of the ability of polar surface to reduce the surface tension of the water, thus allowing water to closely approach the resin surface. In terms of capacity, 0.6 meq/g is the minimum sulfonation necessary to produce a hydrophilic resin surface.

The decrease in $k$ values above 0.6 meq/g may be attributed to lower overall hydrophobicity of the resin at higher concentrations of sulfonic acid groups. The hydrophobic resin matrix may become increasingly shielded by the bulky, polar sulfonic acid groups. All the compounds used in this experiment were quite polar. A similar, although probably not as dramatic, trend would be expected for more nonpolar analytes.

**TABLE 5.3 Comparisons Between Sulfonated (0.4 meq/g) and Unsulfonated Resins**[a]

| | Recovery, % | | | |
|---|---|---|---|---|
| | Sulfonated | | Unsulfonated | |
| Compound | Not Wetted | Wetted | Not Wetted | Wetted |
| Anisole | 94 | 93 | 83 | 89 |
| Benzaldehyde | 90 | 89 | 87 | 96 |
| Nitrobenzene | 96 | 95 | 88 | 96 |
| Hexylacetate | 94 | 94 | 84 | 82 |
| Benzylalcohol | 90 | 98 | 78 | 81 |
| Phenol | 98 | 95 | 77 | 89 |
| Catechol | 59 | 34 | ND[b] | ND |
| *m*-Nitrophenol | 98 | 99 | 89 | 95 |
| Mesityl oxide | 98 | 97 | 93 | 99 |
| 1-2-Hexenylacetate | 93 | 90 | 79 | 89 |
| Average ± RSD %[c] | 95 ± 3.2% | 94 ± 3.4% | 84 ± 5.5% | 91 ± 6.3% |

[a]Wetting solvent is methanol. Values are an average of three trials.

[b]Not detected.

[c]Catechol not included.

*Source:* From reference 7 with permission.

### 5.3.4 SPE with Sulfonated Resins

The ability of sulfonated and unsulfonated resins to extract various organic test compounds from aqueous samples was compared using identical small columns packed with the resins. After the extraction step, the test compounds were eluted with 1.0 ml of ethyl acetate or methanol and determined by GC. The percentage recoveries are given in Table 5.3. The small resin size (8 μm) allows even the hydrophobic underivatized resin to extract the compounds fairly well without methanol pretreatment, but the sulfonated resin, with a more polar surface, is more efficient for extracting these analytes. Note that the sulfonation capacity for this table is 0.4 meq/g, which is close to the optimum capacity of 0.6 meq/g. The ability of water to come into intimate contact with the resin surface facilitates the transfer of analyte from the aqueous sample to the resin surface. The effect of wetting the resin with methanol is also shown. As expected, this has a major effect on the underivatized resin, but is not as important with the sulfonated resin.

## 5.4 RETENTION OF CATIONS BY SULFONATED RESINS; GROUP SEPARATIONS

### 5.4.1 SPE of Bases Using Cation-Exchange Resins

There are a number of older publications dealing with the uptake of protonated organic bases by cation-exchange resins (9–14). A cation-exchange method has been used to concentrate pyridines from beer and worts (15). A comprehensive paper on cation-exchange concentration of basic organic compounds from aqueous solution illustrates this approach (16).

A highly sulfonated resin is used that strongly retains protonated base cations but has little or no affinity for neutral organic compounds. A sulfonated macroporous PS-DVB resin was found to perform better than conventional, gel-type cation exchangers. In the procedure used, Rohm and Haas XAD-4 resin (80–100 mesh) was sulfonated by concentrated sulfuric acid containing some silver sulfate for 2.5 h at 110 °C (16). The capacity of the sulfonated resin, as determined by acid–base titration, was 2.86 meq/g. The resin was used in the $H^+$ form.

A 2.45-g resin sample was packed into a 150 × 8 mm-ID column. The aqueous sample (100 ml–1.0 liter) was passed through the column at a flow rate of about 3.0 ml/min using gravity flow. When all the sample had passed through the column, the reservoir was rinsed twice with 20 ml of pure water. Then the excess liquid was blown out of the column with a gentle stream of air. The column was washed with methanol and with two portions of ethyl ether to remove any neutral analytes adsorbed by the resin.

Desorption of the analyte cations presented some difficulties. The strategy was to convert the analytes from the protonated cationic form to the free base, which can be easily eluted by an organic solvent. However, the large amount of $H^+$- exchange capacity (~7.0 meq) required a large amount of base to neutralize. Initially, either ethyl ether or methanol saturated with ammonia gas was used to neutralize the $H^+$ on the ion-exchange column and elute the uncharged analytes. However, this procedure was lacking in reproducibility owing to the inability to form stable solutions of ammonia in either solvent.

The desorption procedure finally adopted was to neutralize most of the $H^+$ by passing a slow stream of ammonia gas through the ion-exchange column. This reaction could be followed as the resin became lighter brown in color as the $H^+$ resin sites were converted to $NH_4^+$. Then the analytes could be eluted by methanol or ethyl ether containing some $NH_3$. The individual analytes were determined by gas chromatography.

Although cumbersome in some respects, this procedure (16) gave excellent recoveries for a large number of organic bases. The results are summarized in Table 5.4.

The ion-exchange preconcentration method was applied to the analysis of basic organic compounds in oil shale process water. Such samples are vile smelling and contain a wide variety of organic and inorganic substances. After SPE by the cation-exchange procedure, a gas chromatogram (with FID) showed more than 60 peaks of which 6 were identified and quantified (Table 5.5).

**TABLE 5.4 Recovery of Basic Compounds from Aqueous Samples by the Cation-Exchange Method, with Ammonia–Methanol as the Eluting Solvent (16)**

| Sample Compounds | % Recoveries, Range | 50 ppb% Average | % Recoveries, Range | 1 ppm Average |
|---|---|---|---|---|
| 12 aliphatic amines | 72–99 | 86 | 84–99 | 92 |
| 10 aromatic amines | 46–99 | 84 | 48–99 | 91 |
| 27 *N*-heterocyclics | 65–100 | 93 | 71–100 | 96 |

### 5.4.2 Group Separation of Neutral and Basic Compounds

Polystyrenedivinylbenzene beads of high surface area (typically ≈ 400–750 $m^2/g$) are very efficient for SPE of low concentrations of organic solutes in aqueous samples. These resin beads still retain organic solutes when the beads are sulfonated, provided the degree of sulfonation is rather low. The sulfonated beads are able to retain protonated amine cations by an ion-exchange mechanism. So long as the amines are present as cations, they are not washed off the resin by an organic solvent.

These observations are the basis of a method for concentration of analytes from aqueous samples with subsequent separation into neutral and basic groups (17). The scheme is outlined in Figure 5.2. The apparatus used consists of a 30-ml glass reservoir connected by a small adapter to the SPE column itself. Approximately 100 mg of a polymeric cation-exchange resin was packed into a 55 × 6.5-mm-ID column to a height of 12–15 mm. The resin was held in place by 20-μm polyethylene frits placed below and above the resin bed. Flow rate was controlled by air pressure applied to the reservoir.

The resin column is pretreated with 2 ml of concentrated hydrochloric acid and then rinsed with 25 ml of pure water and wetted with a little methanol. This converts the ion exchanger completely to the $H^+$ form. The aqueous sample is acidified with HCl to convert basic analytes to the protonated form. In step 1 the aqueous sample is passed through the resin column at a flow rate of about 1 ml/min. Molecular (neutral) analytes are extracted by the apolar polymer matrix, and the cationic analytes are attracted to the negatively charged sulfonate sites on the resin. In step 2, after a brief wash, the neutral analytes are desorbed and eluted by passing a small volume of methylene chloride through the column. The protonated bases continue to be firmly held by the resin. In step 3 a solution of a volatile

**TABLE 5.5 Compounds Found in Shale Oil Process Water**

| Compound | Concentration, ppm |
|---|---|
| Aniline | 1.3 |
| 4-Picoline | 5.2 |
| 2,4-Lutidine | 17.0 |
| 2,6-Lutidine | 2.9 |
| 2,4,6-Trimethylpyridine | 17.6 |
| Isoquinoline | 1.3 |

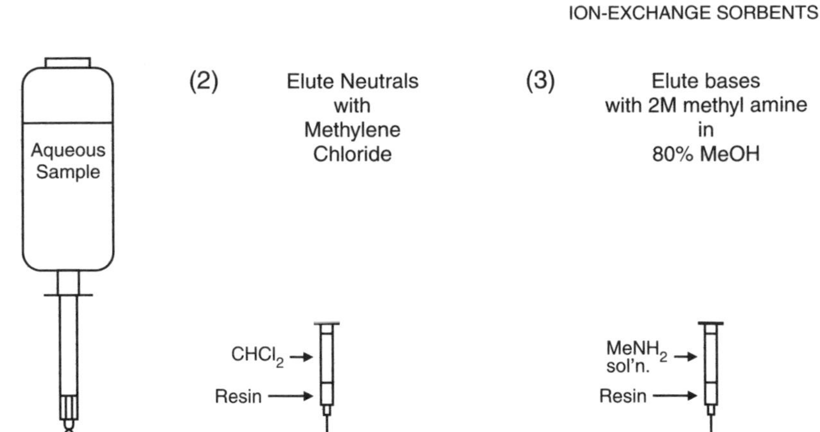


**FIGURE 5.2** Scheme for group separation of bases and neutrals using a cation-exchange resin and two-step elution.

base in methanol is used to elute the basic analytes. The base must be in a sufficiently high concentration to neutralize the $H^+$ sites on the cation exchanger and convert the protonated base ($BH^+$) to the free base (B), which is readily desorbed by the methanol.

***5.4.2.1 Resins*** In order to retain neutral analytes as well as organic cations, the cation exchange resin must be of much lower exchange capacity than that used by Kaczvinsky et al. (16). Schmidt suggested a sulfonate capacity that would facilitate adsorption by both ion-exchange and hydrophobic adsorptive interactions (18). A capacity of 0.6–1.0 meq/g of sulfonate groups was found to be optimum. Lower capacities not only limited the amount of bases that could be retained by an ion-exchange mechanism but also showed lower retention of neutral analytes. Capacities higher than the optimum range showed progressively poorer retention of neutral analytes. The extent of sulfonation, as indicated by the exchange capacity of the sulfonated resin, can be kept low by sulfonation at 0°C for only a short period of time. Successful SPE of both neutral and basic analytes was obtained with two different resins. The first was Rohm and Haas Amberchrome 161 (40-μm spheres with surface area of ~720 $m^2/g$), sulfonated to a capacity of 1.1 meq/g (17). The other resin was a PS-DVB from Sarasep (Santa Clara, CA), ~8 μm spheres with surface area of about 420 $m^2/g$), sulfonated to a capacity of 0.6 meq/g (7).

*5.4.2.2* ***Results*** Aqueous samples containing 5 ppm each of several neutral analytes were passed through a short column packed with macroporous sulfonated resin (1.1 meq/g). After a brief wash with pure water, the analytes were eluted with 1.0 ml of methylene chloride (17). Analysis of a portion of the methylene chloride eluate indicated essentially complete of the test compounds (Table 5.6).

Solid-phase extraction of basic analytes was studied next with the same resin column (17). The results are summarized in Table 5.7. None of the compounds eluted with methylene chloride except 2,3-dimethylquinoxaline, which is an extremely weak base and therefore behaves almost like a neutral compound. Attempted elution of the protonated base cations with 2 M ammonia in methanol gave good recoveries of the weaker bases (pyridine, aniline, etc.) but essentially no elution of the aliphatic amine cations. It appeared that ammonia is too weak a base to convert protonated aliphatic amines to the free base that could then be eluted by the methanol.

**TABLE 5.6 Recoveries of Neutral Organic Solutes**[a]

| Compound | Recovery, % |
|---|---|
| Anisole | 90 |
| Benzonitrile | 95 |
| Benzothiazole | 92 |
| Butylbenzene | 95 |
| Ethylcrotonate | 88 |
| *trans*-2-Hexenylacetate | 96 |
| 2-Hydroxyacetophenone | 99 |
| 4-Nitroacetophenone | 98 |
| Nitrobenzene | 99 |
| 1-Octanol | 94 |
| Octylaldehyde | 97 |
| Octylalcohol | 94 |
| Propylbenzene | 94 |
| Toluene | 98 |
| Average | 95 |

[a]*Conditions:* 10 ml of aqueous sample, pH adjusted to 1.8 with sulfuric acid; cation-exchange resin with 1.1 meq/g sulfonate groups.

**TABLE 5.7 Recoveries of Basic Organic Solutes[a]**

| | Recovery, % | |
|---|---|---|
| Compound | $NH_3$ in MeOH | $MeNH_2$ in MeOH |
| Butylamine | 0 | 103 |
| Cyclohexylamine | 0 | 98 |
| Hexylamine | 0 | 95 |
| Octylamine | 0 | 97 |
| Phenylethylamine | — | 98 |
| Aniline | 95 | 99 |
| *N,N*-Dimethylaniline | 61 | 92 |
| *N*-Methylaniline | — | 92 |
| Pyridine | 91 | 91 |
| Quinaldine | 92 | 95 |
| Quinoline | 92 | 96 |
| Ethylpyridine | — | 93 |
| Isopropylpyridine | — | 101 |
| 2,4-Lutidine | — | 101 |
| 2,4,6-Collidine | — | 95 |
| | Average | 96 |

[a]Conditions were the same as Table 5.5 except for the eluting solvent.

Methylamine in methanol was then tried for desorption of amine cations. Methylamine is a much stronger base (p$K_b$ in water = 3.1) than ammonia (p$K_b$ in water = 4.8). Methylamine is also quite volatile and in the gas chromatographic step elutes well before any of the sample solutes. The last column in Table 5.7 shows excellent recoveries for all of the basic analytes when 2 M methylamine in methanol was used as the eluting solvent.

The effect of pH on group separations was investigated (17). Results were compared for samples buffered at pH 2.0 with 0.1 M sodium dihydrogenphosphate and at pH 7.0 using disodium hydrogenphosphate. At pH 2.0 neutral compounds were eluted with methylene chloride and basic compounds with methylamine in methanol, as expected. Recoveries averaged lower with the pH 2.0 phosphate buffer than when the sample was acidified with HCl. At pH 7.0 the weaker bases (aniline, quinoline, etc.) were not protonated, and therefore were eluted with the neutral group. The stronger bases (hexylamine, ocytlamine, etc.) remained protonated at this pH, and therefore were not eluted until the methylamine in methanol wash step. Thus an additional group separation of aliphatic amines from aromatic amines and nitrogen heterocyclic compounds appears to be feasible.

### 5.4.3 Group Separation of Basic and Neutral Analytes in Organic Sample Solutions

Organic bases dissolved in a nonpolar solvent can be extracted as a group by using a cation-exchange column or membrane (18). An example of this is shown in Figure 5.3 for a toluene solution containing trace amounts of piperidine, 3-picoline, and 2-ethylpyridine. These early eluting bases can-

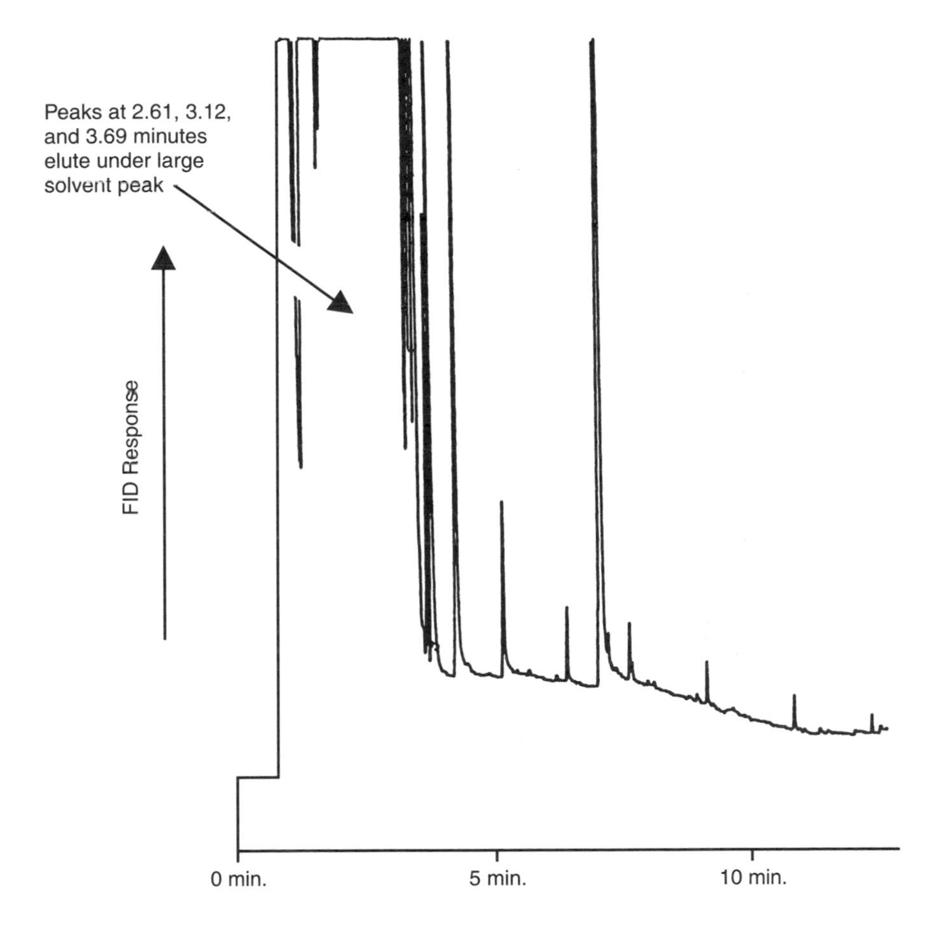


**FIGURE 5.3** Gas chromatogram of toluene containing 6 ppm each of piperdine, 3-picoline, and 2-ethylpyridine (18).

not be separated from the large solvent peak by ordinary GC methods. By adding sufficient HCl ($10^{-4}$ M), the bases are protonated and thus retained by the sulfonic acid groups on the resin. This acid concentration is an optimum and is crucial to the success of the extraction. A higher concentration resulted in increased competition for the cation-exchange sites, while a lower concentration appeared to give incomplete protonation. Both of these scenarios lead to decreased extraction efficiency. After the toluene solution had passed through the membrane, the SPE column was blown dry. The bases were then eluted with 2 M methylamine in acetonitrile and determined by gas chromatography. An example of the resultant chromatogram is shown in Figure 5.4. The average recoveries of four trials was 95–100%.

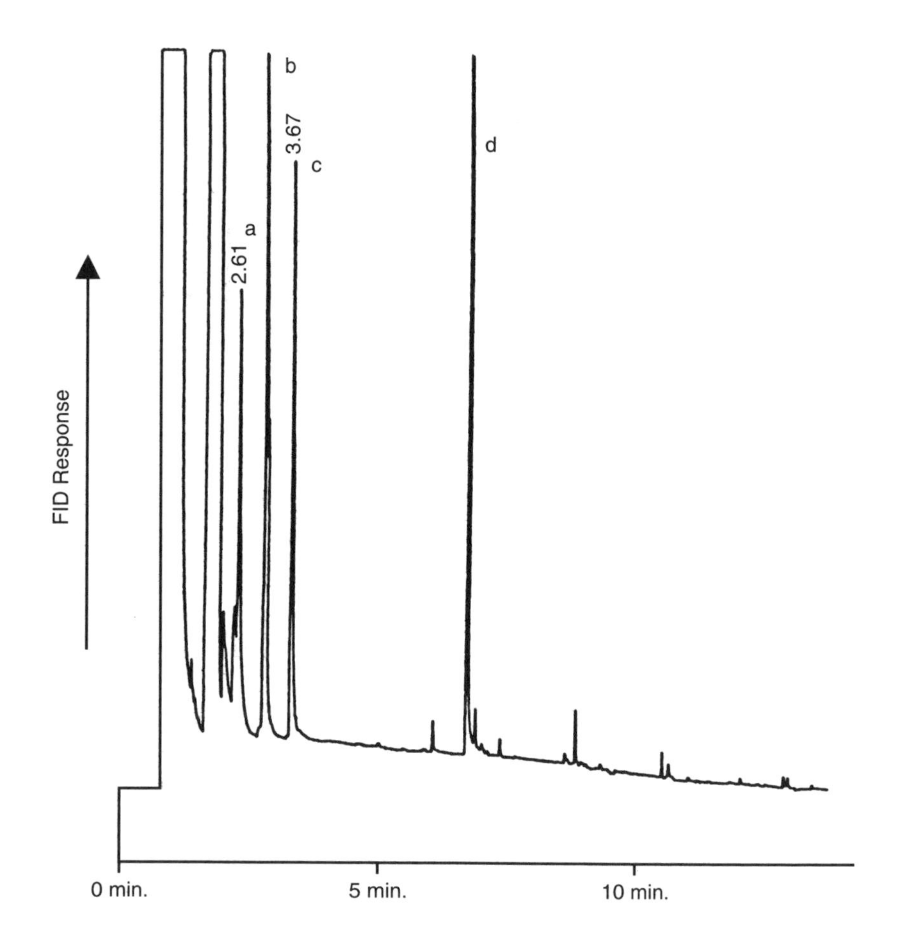


**FIGURE 5.4** Gas chromatogram of bases after extraction from toluene (18) (*a*) piperidine; (*b*) 3-picoline; (*c*) 2-ethylpyridine, (*d*) quinoxaline (internal standard).

This technique appears to be generally applicable for extracting bases present at trace concentrations from complex nonaqueous mixtures. Determination of five bases was attempted in a sample containing benzylamine, phenethylamine, tributylamine, quinoline, and quinaldine, each present at 10 ppm in a toluene solution containing bromobenzene, 3-phenylpropanol, octanol, decanol, dodecanol, and tetradecanol, all at 1500 ppm. Direct analysis of the solution for the bases present by GC was impossible owing to the large peaks present from the various neutrals. But when the basic fraction was extracted by SPE, analysis by GC was possible. The sample solution (2.0 ml) was acidified with HCl (0.0001 M) and passed through the ion-exchange membrane at a flow rate of 1 ml/min. The membrane was then blown dry with air pressure for a period of 1 min. The bases were then eluted with 1 ml of 2 M methylamine in acetonitrile to which 0.1 of quinoxaline was added as an internal standard. Both the basic and neutral fractions were analyzed by GC. Excellent recoveries were obtained.

## 5.5 SPE WITH ANION-EXCHANGE RESINS

### 5.5.1 Retention of Organic Acid Anions

Low concentrations of phenols in water can be isolated in the molecular form by ordinary SPE with polymeric resins (19,20). However, it is desirable to have an extraction method that will concentrate all phenols effectively and permit separation of phenols from other organics that might be present in water. Experiments showed that phenols are effectively retained from water when samples of water containing low concentrations of phenols are rendered basic and passed through a column containing an anion-exchange resin (Rohm and Haas A-26 (21). Exploratory experiments indicated that low concentrations of neutral organic compounds such as naphthalene were only 0–3% retained by the anion-exchange column. Thus, a reasonable selectivity for phenols over other organics was indicated.

The basis of the analytical method reported (21) was as follows. Phenols are taken up as phenolate ions by passing an alkaline water sample through a column of A-26 anion-exchange resin in hydroxyl form. Any neutral organic compounds retained by the resin are removed by washing with alkaline methanol. Phenolate ions continue to be held by the resin during this washing step and are then converted to the molecular form by washing the column with aqueous hydrochloric acid. The phenols are subsequently eluted from the column with acetone–water. The hydrochloric acid and acetone–water effluents are each extracted with methylene chloride. The organic phases are concentrated by evaporation, and the phenols are separated by gas chromatography.

This procedure for SPE of phenols was performed on a fairly large scale. The resin column was 14 × 1.2 cm, and sample volumes as large as 1.0 liter were used. The desorption procedure was rather long and involved. However, excellent recoveries were obtained for low concentrations of phenols in refinery and petrochemical plant effluents and in other water samples.

Some valuable insights into practical aspects of phenol preconcentration are given in Reference 21. As expected, the recovery of phenols was found to be a function of the pH of the sample. In general, phenols are completely retained on the resin if the pH of the sample is at least two units higher than the $pK_a$ of the phenol. All phenols studied were retained at pH between 12.0 and 12.5.

Some very hard water samples contain sufficient bicarbonate to form a copious precipitate of calcium or magnesium bicarbonate when the sample is made basic with sodium hydroxide. This restricts flow through the column. Attempts to avoid precipitation by first adding a complexing agent such as EDTA or tartrate were not effective. The carbonate precipitate can be effectively removed by filtration. This is best done by allowing the precipitate to coagulate for 15–20 min, then filtering it through a sintered glass filter. Paper filters require tedious and lengthy cleaning procedures to avoid introduction of impurities from the paper.

Appreciable amounts of phenols are lost through oxidation in basic solutions and during filtration and column sorption if preventive measures are not taken. Standing at pH 12.5 results in complete loss of phenol within 4 h and *p*-cresol within 24 h. Loss of phenols during slow filtration through filters clogged with calcium carbonate was variable but generally was appreciable, ranging up to 40% of the amount added. No phenols were found adsorbed on the precipitate and thus the losses were attributed to oxidation. When sodium hydrosulfite was added, phenol losses were negligible in basic solutions and were reduced during slow filtrations. Allowing the precipitates to coagulate prior to filtration and decanting the clear water quickly through the filter reduced losses during filtration to undetectable levels.

Many kinds of organic acids can be concentrated selectively and efficiently on anion-exchange columns. Richard and Fritz (22) used an anion resin prepared from a macroporous PS-DVB resin (Rohm and Haas XAD-4) to isolate and concentrate acidic organic material from aqueous solution.

In the procedure developed, water is passed through the anion-exchange column in the hydroxide form, then other organics are removed by washing with methyl alcohol and diethyl ether. Finally, the acids are eluted with diethyl ether saturated with hydrogen chloride gas. The effluent is concentrated, treated with diazomethane, and the methyl esters are separated by gas chromatography using a glass capillary column.

The type of resin used for this kind of SPE is an important consideration. Strong-base resins such as Dowex 1, IRA-400, and A-26 had a tendency to disintegrate on going from an aqueous to a nonaqueous solution and then back to an aqueous medium. The Rohm and Haas XAD resins, however, are highly crosslinked and have excellent and mechanical stability. The resin

**TABLE 5.8 Acids Determined at the 100-ppb Level in Water (22)**

| Sulfonic Acid | Carboxylic Acids | Phenols | Phosphoric Acids |
|---|---|---|---|
| Benzenesulfonic | *o*-Phthalic | 2,4-Dichloro | Dimethyl-phosphoric |
| *p*-Toluenesulfonic | 2,4-Dichloro-phenoxyacetic | 2,4,5-Trichloro | Dimethylthio-phosphoric |
| *n*-Dodecylbenzene-sulfonic | 2,4,5-Trichloro-phenoxyacetic | Pentachloro | Diethylphosphoric |
| *N*-Dodecylsulfonic | 2,Methoxy-3,6-dichlorobenzoic | 2,4,6-Trichloro | |
| 3-Sulfophenyl-butyric | Tetrachloro-terephthalic | 2-Nitro | |
| 3-Sulfophenyl-heptanoic | Acetic | 4-Nitro | |
| *p*-Chlorobenzene-sulfonic | Propionic | 2,4-Dinitro | |
| Sulfoacetic | *iso*-Butyric | 4,6-Dinitro-*o*-cresol | |
| | Butyric | 4-Chloro-3-methyl | |
| | Isovaleric | 2-Chloro | |
| | Valeric | | |
| | *iso*-Caproic | | |
| | Caproic | | |
| | Caprylic | | |
| | Capric | | |
| | Lauric | | |
| | Myristic | | |
| | Palmitic | | |

used was prepared by first chloromethylating XAD-4 and then aminating with trimethylamine to give a quaternary ammonium resin with an exchange capacity of 1.0 meq/g (22).

This anion-exchange method gave excellent recoveries for carboxylic acids, phenols, sulfonic acids, and several other acidic compounds (see Table 5.8). Pesticides in a municipal wastewater treatment plant were determined (22). In a later paper, the method was extended to the determination of organic acids in shale oil water and other complex samples (23).

### 5.5.2 Concentration and Group Separation of Neutral and Acidic Compounds

Earlier, it was shown that partially sulfonated resins have the ability to retain neutral molecules by hydrophobic interactions at unsulfonated areas on the resin surface. This can lead to adsorption in dual-mode consisting of hydrophobic and electrostatic attractions, thus allowing for group separations. In a similar manner anion-exchange resins can be used to perform group separations of acidic and neutral compounds. A scheme for combined concentration and group separation is outlined in Figure 5.5 (24).

Both acidic and neutral compounds are taken up from predominately aqueous samples by an anion-exchange resin column (step 1). Only the neutral compounds are desorbed by washing the resin column with methylene chloride (step 2). Then the acidic sample compounds, which are

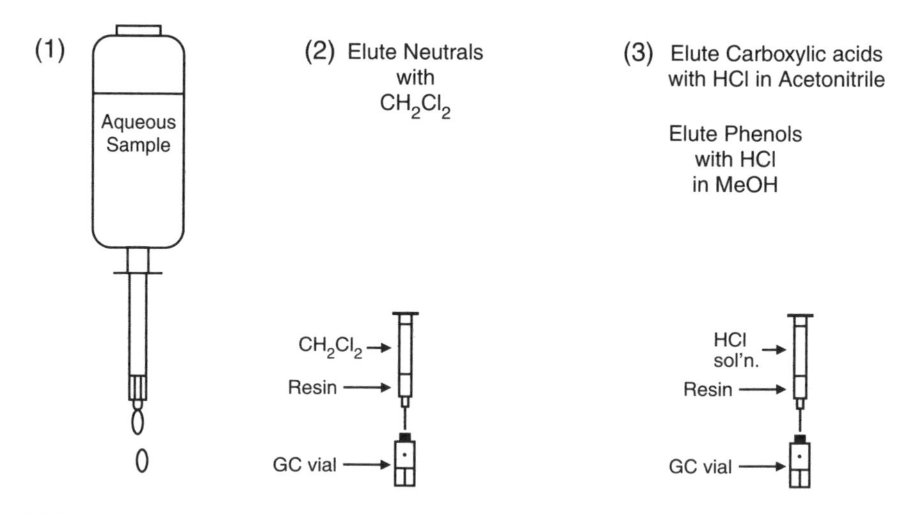


**FIGURE 5.5** Scheme for group separation of neutral and acidic organic analytes.

retained as anions at the resin exchange sites, are desorbed by washing the column with HCl in methanol (step 3).

For this SPE group separation scheme to function properly, the anion-exchange resins must be compatible with organic eluting solvents as well as with the aqueous sample solutions. Macroporous resins work much better than do microporous gel resins. The concentration of polar exchange sites on the resin should be relatively low. Fully functionalized anion-exchange resins do not take up most neutral analytes very well.

In work reported by Schmidt (25), macroporous PS-DVB resins were chloromethylated by a 2.2 M solution of paraformaldehyde in concentrated hydrochloric acid for 24–48 h at 70 °C. Then the resins were aminated by reaction with 25% trimethylamine in ethanol for 24 h. These reactions gave a material with quaternary ammonium groups attached to some of the benzene rings of the polymer: $-CH_2N^+(CH_3)_3Cl^-$.

Two anion exchange resins were prepared (25). One was Amberchrome 161 (Rohm and Haas), 40-μm particles with an exchange capacity of about 0.9 meq/g; the other was Sarasep resin (Santa Clara, CA), 8-μm particles with an exchange capacity of 0.2 meq/g.

For group separations, approximately 100 mg of resin was packed into a small column 6.5 mm ID to a bed height of about 15 mm (for Amberchrome) or about 10 mm for the Sarasep resin (25). The total exchange capacity was much lower than that used in previous methods for SPE of organic anions: about 0.09 meq for Amberchrome and 0.02 meq for Sarasep. After a brief initial cleaning with acetonitrile and methanol, the columns were treated with ~5 ml of dilute sodium hydroxide to ensure that the resin was in hydroxide form. An aqueous sample solution was adjusted to a pH of about 11 with sodium hydroxide and passed through the column at 1 ml/min. After rinsing with 5 ml of pure water, the neutral fraction was eluted with 1.0 ml of methylene chloride. This fraction was collected in a vial, and spiked with quinoxaline or toluene as an internal standard, and the individual analytes were determined by capillary gas chromatography. The acid fraction was eluted from the SPE column with 1.0 ml of 0.1 M HCl in methanol. When the acid fraction contained phenols, the analysis was performed by GC as described for the neutral fraction. Samples containing carboxylic acids were eluted with 1.0 ml of 2 M HCl in methanol and the individual components determined by HPLC.

The eluant used to elute the acid fraction contained hydrochloric acid to convert the sample anions to the molecular form and to neutralize the hydroxide ions remaining on the resin. The molecular analytes could then be eluted by the methanol. It has been shown (J. Chen and J. Fritz,

**TABLE 5.9 Recoveries from 20 ml of Aqueous Sample Using Anion-Exchange Resins (25)**

| Neutral Compounds | % Recovery | Phenols | % Recovery |
|---|---|---|---|
| Anisole | 96 | 4-*sec* Butylphenol | 100 |
| Benzothiazole | 103 | 2-Chlorophenol | 95 |
| Benzylalcohol | 93 | 4-Chlorophenol | 99 |
| Bromobenzene | 90 | 4-Cresol | 98 |
| Chlorobenzene | 87 | 2,5-Dimethylphenol | 94 |
| Cyclohexanol | 100 | 4-Ethylphenol | 100 |
| Decanol | 96 | 4-Isopropylphenol | 100 |
| Ethylbutyrate | 101 | Phenol | 87 |
| Nitrobenzene | 96 | 2-Nitrophenol | 99 |
| Nonylaldehyde | 100 | 3-Nitrophenol | 100 |
| Octanol | 100 | 4-Nitrophenol | 94 |
| Octylaldehyde | 92 | | |
| Salicylaldehyde | 99 | Average | 97 |
| Toluene | 93 | | |
| Triethylorthophosphate | 84 | | |
| Average | 95 | | |

unpublished results) that relatively small concentrations of HCl injected into a GC cause no significant degradation of a capillary column over time.

Table 5.9 shows excellent recoveries for several neutral and phenolic analytes present in aqueous samples in 0.5–5.0 ppm each. Excellent recoveries were also obtained for nine different carboxylic acids present at 5–10 ppm.

Gas chromatograms of the neutral and acid fractions also demonstrate the sharpness of the group separations. For an aqueous sample containing eight neutral analytes and six phenols, Figure 5.6 shows a clean GC separation of the eight neutral compounds (plus quinoxaline as the internal standard) for the neutral fraction. No phenolic peaks are present. In Figure 5.7, the gas chromatogram of the acid fraction shows good resolution of the phenolic peaks.

### 5.5.3 Group Separations in Nonaqueous Solvents

Effective group separations of neutral and acidic can also be carried out when the analytes are dissolved in an organic solvent. For example, a

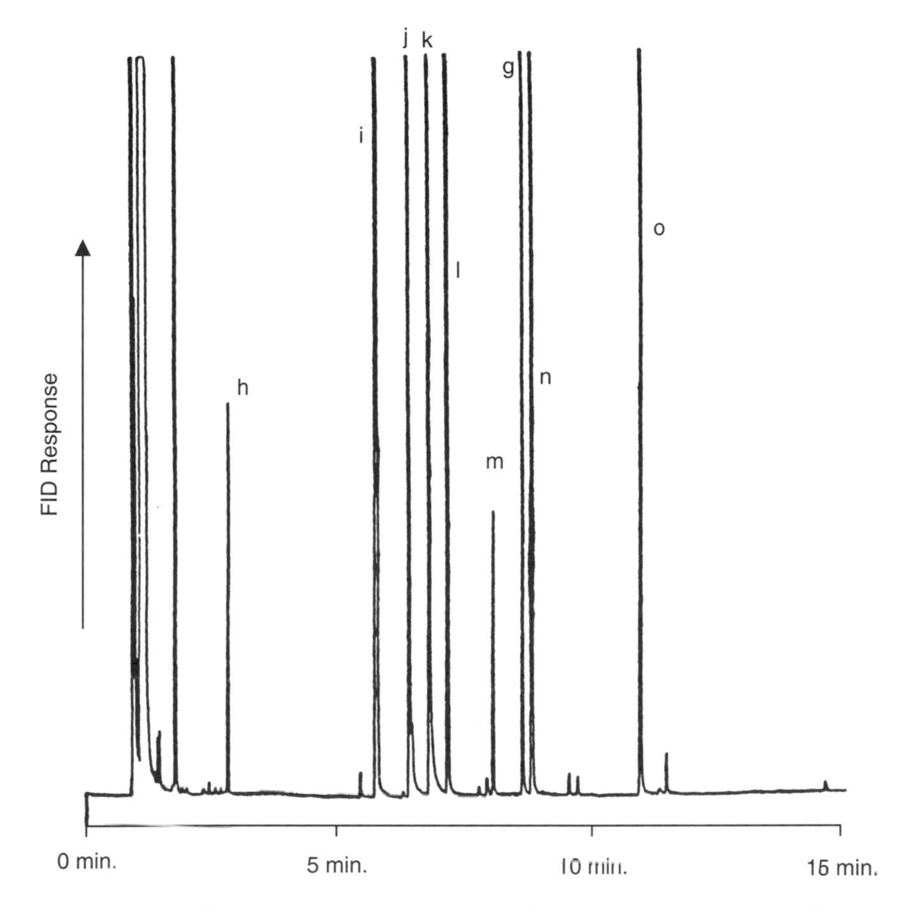


**FIGURE 5.6** Gas chromatogram of neutral fraction concentrated on anion-exchange resin and eluted with methylene chloride: (*h*) toluene; (*i*) benzyl alcohol; (*j*) benzonitrile; (*k*) octylaldehyde, (*l*) nitrobenzene; (*m*) *o*-hydroxyacetophenone, (*n*) benzothiazole; (*o*) *p*-nitroacetophenone; (*g*) quinoxaline (internal standard).

toluene solution containing a mixture of neutral compounds and carboxylic acids was analyzed as follows (26). About 0.1 ml of 1 M methyl amine was added to the sample to convert the acids to anions. This solution was then passed through an SPE column containing ≈10 mg of the 5-μm anion-exchange resin to remove the acids. The neutral solution was then collected and analyzed by gas chromatography. The abstracted acids were then eluted from the column with 1 M HCl in methanol and analyzed by HPLC. Excellent recoveries were obtained for both the neutrals and acids.

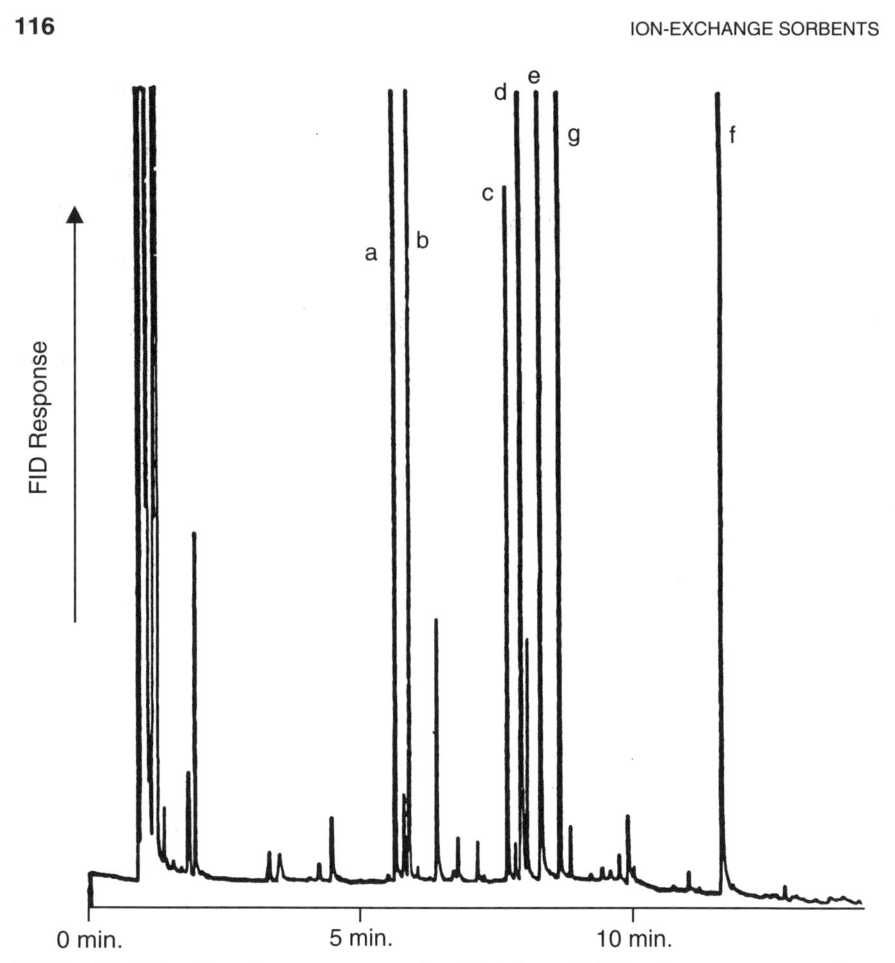


**FIGURE 5.7** Gas chromatogram of acid (phenolic) fraction concentrated on anion-exchange resin and eluted with 0.1 M HCl in methanol: (*a*) phenol; (*b*) 2-chlorophenol; (*c*) 4-chlorophenol; (*d*) *p*-ethylphenol; (*e*) *o*-nitrophenol, (*f*) nitrophenol; (*g*) quinoxaline (internal standard).

## REFERENCES

1. E. M. Thurman and M. S. Mills, *Solid-Phase Extraction*, Wiley, New York, 1998, p. 152.
2. N. Hoffman and J. Liao, *J. Chromamtogr. Sci.* **28** (1990) 428.
3. P. J. Dumont, J. S. Fritz, and L. W. Schmidt, *J. Chromatogr. A* **706** (1995) 109.
4. L. Schmidt and J. S. Fritz, *J. Chromatogr.* **640** (1993) 145.
5. L. W. Schmidt, Ph.D. thesis, Iowa State Univ., Ames, IA, 1993, p. 50ff.
6. J. J. Sun and J. S. Fritz, *J. Chromatogr.* **590** (1992) 197.

7. P. J. Dumont and J. S. Fritz, *J. Chromatogr. A* **691** (1995) 123.
8. M. S. Mills, E. M. Thurman, and M. J. Pederson, *J. Chromatogr. A* **629** (1993) 57.
9. J. J. Richard and J. S. Fritz, *J. Chromatogr. Sci.* **18** (1980) 35.
10. S. R. Watkins and H. F. Watson, *Anal. Chem. Acta* **24** (1961) 334.
11. I. M. Vagina and G. S. Libinson, *Zh. Fiz. Khim.* **41** (1967) 2060.
12. G. S. Libinson and I. M. Vagina, *Zh. Fiz. Khim* **41** (1967) 2933.
13. G. S. Libinson, *Zh. Fiz. Khim.* **45** (1971) 2880.
14. J. Maternova and K. Setinek, *Collect. Czech. Chem. Commun.* **44** (1979) 2338.
15. T. L. Peppard and S. A. Halsey, *J. Chromatogr.* **202** (1980) 271.
16. J. R. Kaczvinsky, Jr., K. Saitoh, and J. S. Fritz, *Anal. Chem.* **55** (1983) 1210.
17. L. Schmidt and J. S. Fritz, *J. Chromatogr.* **640** (1993) 145.
18. L. W. Schmidt, Ph.D. thesis, Iowa State Univ., Ames, IA, 1993, p. 59.
19. G. A. Junk, J. J. Richard, M. D. Grieser, D. Witiak, J. L. Witiak, M. D. Arguello, R. Vick, H. J. Svec, J. S. Fritz, and G. V. Calder, *J. Chromatogr.* **99** (1974) 745.
20. J. A. Vinson, G. A. Burke, B. L. Flager, D. R. Casper, W. A. Nylander, and R. J. Middlemiss, *Env. Lett.* **5** (1973) 199.
21. C. D. Chriswell, R. C. Chang, and J. S. Fritz, *Anal. Chem.* **47** (1975) 1325.
22. J. J. Richard and J. S. Fritz, *J. Chromatogr. Sci.* **18** (1980) 35.
23. J. J. Richard, C. D. Chriswell, and J. S. Fritz, *J. Chromatogr.* **199** (1980) 143.
24. L. W. Schmidt, Ph.D. thesis, Iowa State Univ., Ames, IA, 1993, p. 92.
25. L. W. Schmidt, Ph.D. thesis, Iowa State Univ., Ames, IA, 1993, p. 93ff.
26. L. W. Schmidt, Ph.D. thesis, Iowa State Univ., Ames, IA, 1993, p. 106.

# CHAPTER 6

# RESIN-LOADED MEMBRANES

## 6.1 INTRODUCTION

Cartridges and tubes for SPE are discussed in Chapter 4. A typical cartridge includes a plastic or glass tube with porous metal or plastic frits at both ends. It is filled with 100–500 mg of 40-μm particles. Although SPE cartridges are popular and easy to use, the relatively large particle size of the sorbent and mediocre packing efficiency necessitate the use of a relatively slow flow rate. Small tubes can be packed with smaller sorptive particles and are therefore more efficient for both the extraction and elution steps in SPE. But even with an efficient packed tube, a resin bed height of at least a few millimeters may be necessary to avoid channeling and thus low recovery of extracted compounds.

In 1989 a new family of materials for SPE was introduced (1). These are membranes of polytetrafluoroethylene (PTFE) fibrils impregnated with small particles of solid sorbents such as C18 silica or polystyrene–divinylbenzene. They are manufactured by the 3M Company (St. Paul, MN) under the trade name of *Empore*. Approximately 90% of the weight of the membrane is made up of the sorbent particles. The particles are close together in the membrane but not necessarily touching one another. The resin-impregnated membranes are flexible and generally are ~0.5 mm thick.

A scanning electron micrograph of an Empore membrane is illustrated in Figure 6.1. The sorbent particles are a nonspherical bonded-phase silica with an average particle size of approximately 40 μm. Many of the PTFE

**FIGURE 6.1** Electronmicrograph of a membrane disk. (Courtesy of 3 M Co.)

fibrils that hold the membrane together can also be seen in this photo. The particles within the membrane are held in place to allow the liquid sample to flow evenly through the thin (~0.5-mm-thick) membrane and comes into intimate contact with the extractive particles. It would be very difficult to avoid uneven flow (channeling) through a 0.5-mm bed of *loose* particles packed into a SPE tube.

Flow rates as fast as 200 ml/min through these membranes are possible when used as 47-mm disks in a suction filtration device. Nevertheless, uptake of organic solutes is very efficient because of the fast extraction kinetics. The resin particles are firmly immobilized within the membrane so that channeling does not occur even though the membrane thickness may be $< 1$ mm. Studies with very dilute solutions of a dye show that the dye is taken up by the uppermost part of the membrane. The dye uptake is very even over the entire area of the circular disk.

Other Empore disks contain smaller (average size 8 μm) spherical particles of bonded-phase silica or PS-DVB polymers. Maximum flow rates are somewhat slower than with the larger sorbent particles, but analytes are adsorbed in a thinner, more concentrated layer.

The advantage of a smaller particle size and a dense sorbent bed is shown in Figure 6.2. The smaller particles create a tortuous path that allows analytes to be adsorbed more efficiently.

Elution of sorbed analytes from SPE membranes is also fast and efficient. Owing to the very small bed height in a membrane, the amount of organic solvent needed in the desorption step is generally less than with SPE cartridges or mini columns. Reduction of solvent waste is a very important consideration in the contemporary analytical laboratory.

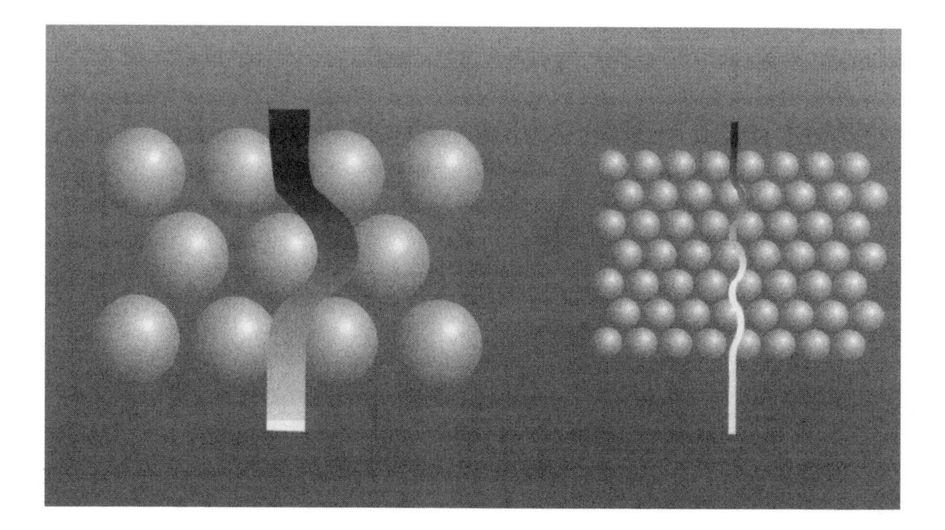

**FIGURE 6.2** Effect of particle size on extraction efficiency of an analyte (dark area) flowing through a resin-loaded membrane. (Courtesy of 3M Co.)

## 6.2 TYPES OF SPE DISKS

Three basic disk types are available commercially for use in solid-phase extraction (2):

1. *Packing-Impregnated PTFE.* These membranes consist of a PTFE fibril network that holds bonded silica particles or resin particles in place. The 8-μm particles constitute ~90% of a device's weight, and the PTFE represents 10%. The disks are available in sizes that fit a standard filter assembly. A 90-mm disk contains ~2000 mg of bonded-phase silica particles, a 47-mm disk contains ~500 mg of sorbent, and a 25-mm disk contains ~140 mg of sorbent. Disks of corresponding diameters filled with styrene–divinylbenzene resins contain approximately one-half the weight of the BP silica disks. These disks are manufactured by the 3M Company and sold by a variety of distributors.

2. *Packing-Impregnated Polyvinyl Chloride.* These microporous plastic sheets contain silica that is activated by a standard ion-exchange or affinity chemistry. The membranes have ~1-μm flow-through pores that provide fast kinetics. Rapid separations can be achieved at low backpressures. The material is available in sheets, cartridges, and disks ranging in diameter from 25 mm to sizes large enough for scale-up work. Flow rates range from 20 to 80 ml/min. Examples of this type of device include Acti-Disk and Acti-Mod products (FMC Corp., Pine Brook, NJ) and Fastchrom membranes (Kontes, Vineland, NJ).

3. *Derivatized Membranes.* These membranes do not contain embedded particles. Instead, these cellulose membranes are functionalized through chemical reactions. Functional groups such as diethylaminoethyl (DEAE) or quaternary ammonium render the membrane an anion exchanger, while a sulfopropyl group allows for uptake of cations. Examples of this membrane type include MemSeps (Millipore Corp., Bedford, MA) and Zeta Chrom devices (CUNO, Meridan, CT).

### 6.2.1 Types of SPE Particles

Empore resin-loaded membranes are available with many of the particles commonly used for liquid chromatography. These include the following:

1. *Bonded-Phase Silica Materials with C18, C8, and C2 Hydrocarbon Chains.* In general, hydrophobic organic analytes are more strongly retained by C18 functionalized particles than by C8.

2. *Styrene–Divinylbenzene (SDB) Resins.* A lightly sulfonated material (SDB-RPS) has a hydrophilic surface that enables the particles to make excellent surface contact with aqueous sample (3,4). These give higher recoveries for phenols, amines, and other moderately polar compounds than most SPE particles. Testing has shown the SDB-RPS absorbent to be effective for explosives, pesticides, phenols, and drugs of abuse. This material can also be used for organic cations, as described in Chapter 3.
3. *Disks with Proprietary Carbon Particles.* These disks have a very high adsorbent capacity. Applications include extraction of oxamyl, a highly water-soluble insecticide that concentrates in the soil, and *N*-nitrosodimethylamine, $(CH_3)_2N{=}N{=}$.
4. *Anion Exchange.* This material is a SDB copolymer functionalized with quaternary ammonium groups. It is used for SPE of anions, such as the routine monitoring of haloacetic acids in drinking water.

Some entirely new types of resin-loaded disks are also becoming available:

1. *Metal-Chelating Resins.* These disks contain chelating polymers that are selective for certain metal ions. One example is an Empore Rad disk containing a chelating material known as *AnaLig* (IBC Advanced Technologies, Inc.), which has a strong affinity for strontium(II). $^{90}Sr^{2+}$ is taken up and can be counted while still on the disk. Another membrane contains embedded particles containing an IDA (iminodiacetic acid) chelating group. IDA is known to form stable chelates with many metal cations. Other Rad disks for radium, technetium (as $TcO_4^-$), and cesium are now available and finding wide use.

## 6.3. TECHNIQUES FOR MEMBRANE SPE

Two major formats are used for SPE with membrane disks. One uses 90-, 47-, or 25-mm (diameter) disks in combination with standard suction filtration equipment. This format is chosen when a large sample is to be used (≥100 ml–1 liter). The other format uses much smaller disks (typically 4–5 mm in diameter) placed in a minicolumn. This method will handle samples of ~1–100 ml and requires a much smaller volume of organic solvent for desorption of sorbed analytes.

### 6.3.1 Procedure for 47-mm Disks

The SPE of phenols from aqueous samples will serve as an example of a typical procedure (5).

For these experiments a 47-mm membrane was placed in a Millipore filtration apparatus attached to a water aspiration. For elution, a 200 × 25-mm test tube was placed in the side-arm flask. Standard samples were prepared by adding several phenols to distilled water so that the concentration of each phenol was approximately 100 μg/liter (0.1 ppm). The pH was adjusted to 2 with hydrochloric acid. In some cases sodium chloride was added to a concentration of 10% w/v.

The disk was placed in the filtration apparatus and washed with acetone to remove any contaminants accumulated from storage, handling, or manufacture. It was then conditioned with a small amount of methanol to promote wetting and uniform flow through the hydrophobic PTFE matrix. The aqueous sample was then passed through the membrane without allowing it to go dry at any time. At full aspirator vacuum the flow rate was approximately 200 ml/min. A typical sample volume was 500 ml. The retained phenols were eluted with three 3-ml portions of tetrahydrofuran. Each portion was allowed to soak into the disk for about 5 min before pulling it through. The combined aliquots of tetrahydrofuran were made up to exactly 10 ml.

### 6.3.2 Procedure for Small Disks in a Tube

Extraction of phenols from aqueous samples will again serve as an example of a typical procedure (5).

These experiments were carried out with a small column obtained from Alltech (Deerfield, IL). A typical tube size was 55 × 7 mm. Membrane circles were cut slightly larger in diameter than the tube. These were forced into the tube with 20-μm polyethylene frits above and below for support. Six circles of membranes were placed in the tube, giving a bed height of ~3 mm. The SPE tube was then connected to a laboratory-made reservoir. Sample solutions were then forced through the membrane at a controlled rate by air pressure applied at the top or by suction from below.

The sample solution was prepared by adding a standard methanol solution of phenols (50 ppm each) to 20 ml of water so that the final concentration was approximately 0.4 ppm each. The pH was adjusted to approximately 2 with sulfuric acid to repress the ionization of the more acidic phenols.

Prior to each use a small amount of methanol (~1 ml) was added to wet the disk. Without allowing the column to go dry, the aqueous sample solution was passed through at a flow rate of 2 ml/min by applying a pressure of 15 psi (103 *k*Pa). The column was then washed with 2 ml of distilled water.

The phenols were eluted with 0.75 ml of methanol. The eluate was then collected in a 1.9-ml GC vial, and 0.1 ml of internal standard (500 ppm toluene in methanol) was added. The vial was then capped and a 2-μl aliquot was injected onto the gas chromatograph. The percentage recovery was calculated by comparing the peak heights (relative to the internal standard) with those of the original standard methanol solution of the phenols. Internal standard was also added to the methanol solution.

Circles can be cut from larger membrane disks by means of a cork bore. It is necessary to cut the circles slightly larger than the inside diameter of the tube to ensure a very snug fit. If this is not done, the sample solution may flow preferentially along the tube walls with poor extraction. In later experiments, a single membrane circle ~3 mm thick was used with success.

### 6.3.3 Extraction Disk Cartridges

Empore high-performance disk cartridges are perhaps the most convenient and efficient devices available for SPE on a relatively small scale. Particle-loaded membranes are mounted into standard 1-, 3-, or 6-ml polypropylene syringe barrels, with effective diameters of 4, 7, and 10 mm, respectively. The membrane disks are about 0.5 mm thick and rest on a support that is designed to minimize dead volume. The membrane disk is held in place by a top ring. An efficient prefilter on top of the membrane reduces the possibility of plugging with biological fluids or sediment. The prefilter also provides consistent flow properties. All components of this device are made from ultrapure polypropylene except for the membrane, which contains PTFE fibrils and sorbent particles.

Volumes of conditioning solvent, sample, wash solution, and elution solvent are summarized in Table 6.1. The general procedure for SPE with disk cartridges is as follows:

1. Condition the disk (usually with methanol). This step is necessary to ensure good contact between the aqueous sample and the membrane surface. Use vacuum to pass most of the methanol, then

**TABLE 6.1 Volume Guidelines for SPE Using Disk Cartridges**

| | Disk Diameter | | |
|---|---|---|---|
| Step | 4 mm | 7 mm | 10 mm |
| 1. Condition | | | |
| Methanol | 0.15 ml | 0.25 ml | 0.50 ml |
| Water | 0.25 ml | 0.50 ml | 0.75 ml |
| 2. Load | | | |
| Sample | 0.1 ml | 1.0 ml | 2.0 ml |
| Buffer | 0.1 ml | 1.0 ml | 2.0 ml |
| 3. Wash | 0.15 ml | 0.25 ml | 0.50 ml |
| 4. Elute | 150 μl | 300 μl | 500 μl |

remove the residual methanol with distilled water. Use vacuum to pass most of the water, but leave the surface of the disk wet.

2. Extract the sample. Carefully transfer the sample into the extraction cartridge. Adjust the pH if the method requires and add an internal standard if one is needed. Pass the sample through the disk with vacuum. The disk should not be allowed to become dry prior to sample addition or during the sample extraction step.
3. Wash. Use a wash with distilled water, buffer, or an organic–aqueous solvent mixture to remove interferences while leaving analytes retained on the disk.
4. Elute. Using a clean sample collection vial or tube, add a suitable volume of eluting solvent and pass this through the disk using a gentle vacuum. If necessary, turn off the vacuum and add a second portion of eluting solvent.

## 6.4 SEMIMICRODISK EXTRACTION

Membranes require less eluting solvent and are generally more efficient than cartridges and tubes containing loose sorbent particles. However, only a small fraction of the eluted sample (1–2 μl) is generally used for measurement of analytes by gas chromatography. Large membrane disks require up to 10 ml of eluting solvent. An injection of 2 μl of the eluate into a gas chromatograph represents only 2/10,000 or 0.02% of the sample analytes.

A 7-mm-diameter disk cartridge may require 300 μl of eluting solvent (Table 6.1). A 2-μl injection of this eluate represents 2/300 or 0.67% of the sample analytes.

A higher percentage of the sample components will be used for the chromatographic analysis when a smaller volume of eluting solvent is used. In Chapter 8 a semimicroscale device for SPE is described in which a small membrane disk is used. Only 30–50 μl of eluting solvent is needed (6). Thus an average of 2/40 or 5.0% of the sample analytes is injected. This is 5/0.67 or 7.5 times greater than the percentage calculated above for the disk cartridge. Since the height (or area) of a chromatographic peak depends on the total amount of sample component injection, this means that the detection limit will be lower when a higher percentage of the eluate is injected.

Another way to increase the percentage of the sample analytes used for chromatography is to evaporate the volume of eluate to a smaller volume. This technique was used extensively in the early days of SPE (Chapter 2). While evaporation gives lower limits of detection, it reduces the speed of SPE and risks the chance of some loss of sample analytes.

## 6.5 RECOVERY OF TEST COMPOUNDS USING MEMBRANE DISKS

### 6.5.1 Comparison of SPE Sorbents

The efficiency of SPE of a number of test compounds was compared using Empore membranes containing C18 silica, underivatized PS-DVB, and an experimental disk containing acetyl-derivatized PS-DVB particles. Six layers of each membrane (~0.5 mm thick) were packed into a tube 4.5 mm ID. After conditioning with methanol, aqueous samples (20 ml) containing about 0.4 ppm of each test compound were passed through the tube containing the membranes at a rate of 2 ml/min. After a brief wash (2 ml of water), the extracted analytes were eluted with 0.75 ml of ethyl acetate and the individual analytes measured by gas chromatography. The recoveries, which averaged four individual runs, are given in Table 6.2 (5). These results show that for many kinds of test compound recoveries are significantly higher with polymeric resins than with silica C18. The presence of hydrophilic acetyl groups on the polymeric resin was also shown to give better recoveries for phenols than the underivatized PS-DVB resins.

**TABLE 6.2 Recoveries of Organic Solutes (1 ppm) with a Sulfonated Empore Membrane (0.6 meq/g)**

| Compound | Recovery, % |
|---|---|
| Phenol | 98 |
| *p*-Cresol | 102 |
| 2,5-Dimethylphenol | 98 |
| *p*-Chlorophenol | 97 |
| *o*-Chlorophenol | 95 |
| *m*-Nitrophenol | 97 |
| 4 *sec*-butylphenol | 98 |
| 4-*tert*-butylphenol | 100 |
| 4-Hexylresorcinol | 101 |
| 2-Methylresourcinol | 91 |
| *p*-Isopropylphenol | 96 |
| Dodecylalcohol | 94 |
| 1-Hexanol | 94 |
| Cyclohexanol | 93 |
| 2-Ethyl-1-hexanol | 97 |
| Benzylalcohol | 94 |
| 1-Octanol | 96 |
| Phenethylalcohol | 96 |
| 3-Phenyl-1-propanol | 98 |
| Benzonitrile | 99 |
| Nitrobenzene | 100 |
| 3-Nitroacetophenone | 98 |
| Benzothiazole | 97 |
| *o*-Nitrotoluene | 97 |
| Isophorone | 100 |
| Benzophenone | 96 |
| Acetophenone | 102 |
| Hexylaldehyde | 95 |
| Octylaldehyde | 96 |
| Nonylaldehyde | 93 |
| 9-Anthraldehyde | 98 |
| Salicylaldehyde | 100 |
| Benzaldehyde | 104 |
| Anisole | 96 |
| Phenetole | 91 |
| Ethylacetoacetate | 97 |

(*cont.*)

**TABLE 6.2** (*Continued*)

| Compound | Recovery, % |
|---|---|
| Methylbenzoate | 96 |
| Ethylcinnamate | 95 |
| Hexylacetate | 93 |
| *tert*-2-Hexenylacetate | 91 |
| Isopentylbenzoate | 92 |
| 1-Iodoheptane | 98 |
| 1-Bromododecane | 92 |
| 1-Chlorododecane | 90 |
| Average ± RSD (%) | 96 ± 3.1 |

### 6.5.2 SPE with Lightly Sulfonated PS-DVB Resins

In Chapter 5 the ability of sulfonated and unsulfonated resins to extract various organic test compounds from aqueous samples was compared using identical small columns packed with the resins (4). After the extraction step, the test compounds were eluted with 1.0 ml of ethyl acetate or methanol and determined by GC. The small resin size (8 μm) allows even the hydrophobic underivatized resin to extract the compounds, but the sulfonated resin, with a more polar surface, is even more efficient for extracting these analytes. The effect of wetting the resin with methanol had a major effect on the underivatized resin, but is not as important with the sulfonated resin. The surface of the sulfonated resin is hydrophilic enough that methanol does not significantly modify it.

Empore membranes embedded with sulfonated resin of approximately 0.6 meq/g were also used for SPE (4). These membranes embedded with other materials have been used and described previously (1,2,5). They offer several advantages over loose resin, including lower backpressure necessary to load samples, decreased channeling, and improved mass transfer (7,8). In this study, sulfonated membranes were used to extract neutral organic compounds from water. Averaged triplicate recoveries of 45 analytes are shown in Table 6.2. Many classes of compounds are represented, including phenols, alcohols, aldehydes, ketones, and esters. Polar analytes, like the phenols, and nonpolar analytes such as the halogenated alkanes are all efficiently recovered. Recoveries were over 90% for all compounds.

### 6.5.3 Silicalite Disks

In the procedures described in Chapter 3, particles of a molecular sieve known as Silicalite were used for SPE or organic analytes from aqueous samples. Small hydrophilic compounds, such as the lower alcohols, aldehydes, ketones, and esters, are well extracted by Silicalite, thus adding a valuable new dimension to conventional SPE. Some more hydrophobic molecules are also extracted, but large bulky molecules are excluded from the 6-Å channels and are therefore poorly extracted.

**TABLE 6.3 Comparison of Percentage Recoveries of Various Analytes Using Silicalite Particles and a Silicalite-Loaded Membrane**

| Class | Compound | Particles | Membrane |
|---|---|---|---|
| Alcohols | Methanol | 7 | 1 |
| | Ethanol | 54 | 5 |
| | 1-Propanol | 84 | 86 |
| | 1-Butanol | 95 | 98 |
| | 2-Butanol | 94 | 98 |
| | *tert*-Butanol | 95 | 100 |
| | 1-Pentanol | 92 | 100 |
| | 2-Pentanol | 94 | 99 |
| | 1-Hexanol | 99 | 99 |
| | Cyclohexanol | 10 | 53 |
| | 1-Octanol | 89 | 89 |
| | 1-Decanol | 82 | 78 |
| | 1-Dodecanol | 75 | 73 |
| | 1-Tetradecanol | 58 | 57 |
| | 3-Phenyl-1-propanol | 73 | 89 |
| | 2,2-Dimethyl-3-pentanone | 94 | 98 |
| | 2-Ethyl-1-hexanol | 92 | 98 |
| Esters | Methylacetate | 85 | 4 |
| | Ethylformate | 83 | 99 |
| | Ethylacetate | 91 | 99 |
| Carboxylic acids | Acetic acid | 2 | 0 |
| | Propionic acid | 62 | 5 |
| | Butyric acid | 78 | 84 |
| | Valeric acid | 95 | 100 |
| Phenols | *p*-Cresol | 100 | 94 |
| | *o*-Cresol | 97 | 92 |
| | 4-Isopropylphenol | 89 | 89 |

Silicalite particles have also been incorporated into an experimental (not commercially available) membrane of the Empore type. Several circles were cut from the membrane and packed into a small column to the same height as that previously packed with loose Silicalite particles. Results of SPE with the loose particles and membrane column were then compared. The results in Table 6.3 showed much lower recoveries of ethanol and methyl acetate with the membrane compared to the loose particles (6,9). On the other hand, the recovery of cyclohexanol was higher with the membrane column. Excluding methanol, ethanol, cyclohexanol, methylacetate, acetic acid, and propionic acid, the average recovery was 84% with both the membrane and the loose Silicalite.

## 6.6 APPLICATIONS

Solid-phase extraction membranes (SPEMs) are used to facilitate the analysis of a wide variety of "real-world" samples. The nature of these samples often presents special problems that require modification of the basic SPE procedure. In this section we discuss several examples that illustrate how these problems have been solved. These applications will also serve to give a feeling for the scope of SPE with membranes.

### 6.6.1 Biological Samples

Lensmeyer et al. devised a therapeutic drug monitoring procedure using a SPEM for isolation of antiarrhythmic drugs from serum (10). Commercial Empore C18, 47-mm-diameter membranes were cut into individual 11-mm disks using a cork borer, and these were secured in a MF-1 microfilter holder. In their basic procedure 500 μl of serum and 1.0 ml of an internal standard solution were combined into a polypropylene microfuge tube (1.7 ml size) and certrifuged for 3 min at 12,400*g* to sediment particulate matter that might plug the extraction membrane. The liquid was carefully transferred to the microfilter holder assembly. The MF-1 assembly was centrifuged (85*g*, swinging-bucket type) for 10 min to force the sample through the membrane. After a wash, a clean receiving tube was attached; 350 μl of HPLC mobile phase was added to the reservoir and the unit centrifuged for 5 min to desorb the extracted drugs.

A similar procedure was used to extract steroids from serum samples prior to their separation by HPLC (11). In this case the serum sample was incubated at room temperature for 10 min to release steroids bound to proteins. These procedures demonstrate the effectiveness of SPEM in concentrating drugs without additional manipulations.

Extraction and quantification of organic compounds from biological samples such as urine and serum present several specific problems. Biological samples contain components such as proteins, lipids, saccharides, and salts in varying concentrations. The organic compounds to be extracted vary in polarity and may associate with the matrix components. The fact that some of the large matrix molecules in urine and serum may plug up the SPE membrane is of even greater concern.

Experiments showed that an experimental membrane containing Silicalite is more resistant to plugging than are the other membranes tested (9). However, diluted urine samples still required the addition of 10% v/v of methanol, or better, addition of 10% v/v of 30 mM aqueous sodium dodecylsulfate (SDS) to reduce the backpressure. SDS binds to proteinaceous material and releases protein-bound drugs and other agents (12,13). As a result, proteins are not adsorbed onto the membrane and thus do not interfere with the SPE process. The chromatograms of eluates from samples treated with SDS were much cleaner than those containing methanol. The average recovery of all compounds in SDS-treated urine samples was 88% with a relative standard deviation of 2.5%.

Serum samples required dilution (1:2) using an aqueous 30 mM solution of SDS. The serum samples were more viscous than the aqueous–albumin samples and contained more potential interferences. After many extractions, with the same membrane the backpressure did not increase, implying no fouling of the membrane by protein.

Fresh urine and human serum samples were analyzed to determine the extraction ability for samples of this type. Compounds previously identified in human urine (14,15) and serum (14,16,17) were added to the samples in low concentrations. An experimental Silicalite-loaded membrane was used for these experiments. Recovery data for the test compounds are given in Table 6.4. The average recovery of all compounds in SDS-treated urine samples was 88% (RSD 2.5%) and the average recovery for 18 test compounds in serum was 91%.

### 6.6.2 Water Samples Containing Particulates

The rapid sample flow rates that can be achieved with SPEMs means that sample preparation time is significantly reduced compared to cartridge SPE. These improved flow rates make the disks attractive for the analysis of particle-laden water samples. However, there have been two major con-

**TABLE 6.4 Recoveries of Test Compounds from Various Samples Using a Silicalite-Loaded Membrane**

| | Recovery, % | |
|---|---|---|
| Compound | Urine, SDS | Serum, SDS |
| 1-Butanol | 95 | 90 |
| 2-Butanol | — | 87 |
| 1-Pentanol | 89 | 101 |
| 2-Pentanol | 91 | — |
| 1-Hexanol | — | 101 |
| 2-Ethyl-1-hexanol | — | 93 |
| 1-Octanol | 93 | 100 |
| 2-Octanol | — | 100 |
| 2-Butanone | 100 | 100 |
| 2-Pentanone | 100 | 100 |
| 4-Methyl-2-pentanone | 97 | 88 |
| 2-Hexanone | 100 | 99 |
| 3-Hexanone | 88 | — |
| 2-Heptanone | 94 | 100 |
| 2-Methylbutyraldehyde | 59 | — |
| *n*-Valeraldehyde | — | 90 |
| *trans*-2-Pentenal | 93 | — |
| Hexaldehyde | — | 97 |
| Benzaldehyde | 99 | 97 |
| Thiophene | 81 | — |
| *R*-Carvone | 92 | — |
| Benzene | 78 | 55 |
| Toluene | 65 | 76 |
| Phenol | 88 | — |
| *p*-Cresol | 100 | — |
| Chloroform | 58 | 68 |

cerns: (1) the suspended solids in the water sample can plug the SPE sorbent bed, and (2) the analytes of interest may be adsorbed onto the particulates. Concern also exists as to whether small particulates can pass through the pores of the sorbent bed, resulting in loss of the analyte.

The worst plugging problems are most often encountered with surface waters that are high in biological activity or with water containing fine, suspended clay particles. When the sorbent bed becomes plugged with particulates, sample flow slows down until it effectively stops.

The severity of a plugging problem depends on the concentration, type, and size of the particulates in the sample, the pore size of the sorbent, and the surface area of the sorbent bed. The smaller particle size of sorbent particles used in SPE disks (10 μm) versus SPE cartridges (40 μm), as well as the larger surface area of the disks, makes them less susceptible to plugging than cartridges.

After examining several approaches, Dirkson et al. found that small, high-density glass beads, used as in situ filtration aid in conjunction with a 90-mm SPEM, was very effective in maintaining high flow rates (18). A 12.5-mm layer of the glass beads was placed on top of the SPE disk. Flow rates for several types of sediment-containing water samples were much higher with the glass beads. The glass beads were also found to be effective in preventing sediment with adsorbed analyte from passing through the SPEM. As a result, significantly higher recoveries were obtained on sediment-laden samples when glass beads were used as filter aid.

### 6.6.3 Pesticides

The determination of low concentrations of pesticides in natural waters and in drinking water is essential in evaluating environmental problems and in ensuring the purity of potable water. Liquid–liquid extraction was formerly used to concentrate pesticides prior to their analytical determination, but more recently SPE and especially SPE with resin-loaded membranes are coming into wider use. Font et al. (19) published a detailed review with 129 references on the use of solid-phase extraction in multiresidue pesticide analysis of water. They point out that a new generation of SPE devices has emerged that include flat disks with large cross-sectional areas. These provide advantages for on-line preparation and clean-up methods with respect to sorption, capacity, back pressure and stability after repeated use (1,20–23).

The two vital steps in any preliminary step involving extraction are (1) effective extraction of the sample pesticides and (2) complete desorption of the extracted substances prior to analysis. The first (extraction) step depends not only on the use of an effective solid-phase extractant but also on the nature of the water sample. Significant losses in recovery tests on pesticides have been observed with SPE when water samples with high contents of organic matter have been analyzed. This stems from competition for the active sites of the adsorbent between the chlorinated hydrocarbons and other hydrophobic groups present in the sample.

On analyses of marine and surface waters containing solid particles forming suspensions, the recoveries from unfiltered waters were found to be substantially lower than expected for some pesticides (24). Humic substances in water can increase the apparent solubility of these compounds, and bind organic compounds with covalent bonds, as charge-transfer complexes, by hydrogen bonding or by van der Waals interactions. These substances are adsorbed on the suspended solid particles (25–27).

Desorption of adsorbed sample compounds is generally accomplished by washing the sorbent particles with a small volume of an organic liquid or a mixture of liquid solvents. The use of a small SPEM generally requires a smaller elution volume than do cartridges packed with loose sorbent particles.

Thermal desorption (24) has also been applied for the desorption of some organochlorine pesticides. The extraction column is placed inside the GC oven. Thermal desorption can fail as a result of a very strong interaction between the pesticide and the solid sorbent. The high temperature needed for desorption may decompose the pesticide molecule.

Several authors have used supercritical fluid extraction to remove pesticides from the solid sorbent (28–30). One advantage of this approach is that the supercritical fluid (SF) is easily evaporated after the desorption step, leaving a small, concentrated residue for further analysis. However, SFE has its downside. Supercritical $CO_2$ is often a rather poor solvent for the desorption step and may require the use of a conventional liquid solvent as an additive. This additive solvent, as well as residual water in the SPE device, may cause plugging of the restrictor (needed to maintain pressure needed for a SF) by ice formation.

Still another possibility is to determine extracted sample compounds while they are still on the SPEM. Poziomek et al. suspended membrane tabs in solutions containing ppb concentrations of anthracene for specific periods of time. The tabs were then withdrawn and allowed to dry, and the anthracene was measured by nondestructive fluorescence spectroscopy (31).

### 6.6.4 Oil and Grease

The determination of oil and grease in a wide variety of food and industrial samples is a mundane yet vital analytical determination. In EPA Method 413.1 a 1-liter aqueous sample is adjusted to pH 2 and extracted (liquid–liquid extraction) 3 times with Freon 113. The Freon is then evaporated and the oil and grcase residue is weighed. In an effort to speed up the method and also eliminate the use of toxic Freon, a procedure was developed in which the sample is filtered through a 90-mm C18 disk. The adsorbed oil and grease is then eluted with hexane after an initial preelution conditioning with acetone to make the water miscible (32). This method does work for oil and grease, but the quantitative results were not always in agreement with EPA Method 413.1.

### 6.6.5 Membranes Containing Ion-Exchange Resins

In Chapter 5 the use of cation- or anion-exchange materials as solid-phase extractants was discussed. A sulfonated material with a high exchange capacity (2–4 mmol/g) will retain cations by a cation-exchange mechanism. Owing to the high degree of sulfonation, neutral organic compounds will not be retained or will be weakly retained and easily washed through the resin column. Similarly, anion exchangers with a high concentration of quaternary ammonium groups will take up anions but will retain neutral compounds weakly if at all.

Ion exchangers with a much lower concentration of exchange groups behave differently. For example, a macroporous PS-DVB resin containing 0.5–1.0 mmol/g of $-SO_3^- H^+$ groups will retain cations by an ion-exchange mechanism but will also take up neutral organics by an adsorption mechanism. Neutral compounds are eluted efficiently by an organic solvent while the cationic analytes must be converted into neutral compounds (by pH adjustment) before they can be eluted. This was the basis of a group separation of neutral and basic compounds described in Chapter 5. In a similar manner separation of neutral and acidic groups can be accomplished by using a polymeric anion exchanger containing only a low concentration of quaternary ammonium groups.

SPEMs loaded with appropriate ion-exchange resins are more efficient for these separations than are loose resins packed in a tube. For example, Fritz, Dumont, and Schmidt described a concentration/separation procedure in which a macroporous anion-exchange resin of fairly low exchange capacity was used (33). The resin tube was treated with dilute sodium hydroxide to convert it to the $OH^-$ form. The aqueous sample was also rendered basic to convert the acidic organic solutes to the anionic form. After passing an aqueous sample through the resin column, neutral solutes were eluted with 1 ml of methylene chloride and measured by GC. Then the acidic substances were eluted by 1 M HCl in acetonitrile or methanol. Excellent group separations of neutral/acidic compounds were obtained, and the overall recovery of test compounds was also very high (Table 6.5).

**TABLE 6.5 Average Recoveries of Neutral and Acidic Organic Compounds Concentrated from Dilute Aqueous Solution by Anion-Exchange Resin**

| Neutral Compound | Recovery, % | Acidic Compound | Recovery, % |
|---|---|---|---|
| Ethylbutyrate | 101 | *p*-Hydroxybenzoic acid | 99 |
| Chlorobenzene | 87 | Benzoic acid | 96 |
| Cyclohexanol | 100 | *p*-Nitrobenzoic acid | 99 |
| 1-Bromohexane | 90 | 1,2,4-Benzenetricarboxylic acid | 105 |
| Salicylaldehyde | 99 | Salicylhydroxamic acid | 102 |
| 1-Octanol | 100 | 2,4-Dihydroxybenzoic acid | 99 |
| Nonylaldehyde | 100 | Isophthalic acid | 100 |
| Triethyl orthopropionate | 84 | Phenol | 87 |
| 1-Decanol | 96 | 2-Chlorophenol | 95 |
| Benzylalcohol | 93 | 4-Chlorophenol | 99 |
| Octylaldehyde | 92 | 2-Nitrophenol | 99 |
| Nitrobenzene | 103 | 3-Nitrophenol | 100 |
| Toluene | 93 | *p*-Cresol | 98 |
| Anisole | 96 | 2,5-Dimethylphenol | 94 |
| | | 4-Isopropylphenol | 100 |

Neutral compounds were eluted with methylene chloride and acidic compounds were subsequently eluted with 1 M HCl in methanol.

To comply with anticipated regulations, drinking water must be monitored routinely for haloacetic acids, which are byproducts of some disinfection methods. EPA Method 552.1 concerns a SPE procedure in which a resin cartridge is used to extract and concentrate haloacids as well as the chlorinated herbicide Dalapon. However, Empore anion-exchange extraction disks (47 mm) are also accepted for use in Method 552.1. The 100-ml water samples can be filtered through the extraction disks at full vacuum in as little as 2.0 min and still give high extraction efficiency.

### 6.6.6 Metal Ions

It is frequently necessary to isolate low concentrations of metal ions from samples containing high concentrations of salt or other commonly encountered ions. Sometimes it is also necessary to preconcentrate the selected metal ions to bring their concentration into a range suitable for measurement. Preconcentrations of metal ions can be accomplished with solid particles that contain functional groups that will form chelates with the desired metal ions. An even better procedure is to use a membrane loaded with an effective chelating resin. SPE methods for metal ions will be discussed in Chapter 7.

## 6.7 CONCLUSIONS

At this point it might be well to summarize the relative advantages of using resin-loaded membranes or loose solid particles packed in cartridges or tubes. Packed SPE devices often have a particle bed several centimeters in height. This provides a relatively high capacity for extraction. The high bed height may also give better retention of analytes with low capacity factors that might be incompletely retained in a SPE device with a smaller bed height. On the other hand, a relatively high bed height means that a higher volume of desorbing solvent will be needed. Pollution from organic solvents and disposal of liquid organic waste is becoming a serious concern. The volume of desorbing solvent needed can often be reduced by reversing the direction of flow so that analytes sorbed at the top of the column are first to be eluted. However, it is not always convenient to reverse the flow for the elution step.

By immobilizing sorbent particles within a membrane, a bed height of $\leq$ 0.5 mm can be used without sample loss due to channeling. Smaller particles ($\leq$5 μm) can be used in membranes while still maintaining a reasonable flow rate through the membrane. Large samples can be filtered rapidly through a 47- or 90-mm membrane disk while smaller samples can be used in conjunction with disks only 4–5 mm in diameter. Excellent retention of sample components is obtained even on thin disks. The volume of eluting solvent used with membranes is generally much lower than with cartridges or minicolumns packed with loose particles. Thus, evaporation of effluent prior to analysis is reduced or even eliminated completely. Generation of liquid waste is significantly reduced.

On balance, the better efficiency, speed, and lower waste volumes point to increasing use of membranes for analytical solid-phase extraction.

## REFERENCES

1. D. F. Hagen, C. G. Markell, G. A. Schmitt, and D. D. Blevins, *Anal. Chim. Acta* **236** (1990) 157.

2. C. Markell, D. F. Hagen, and V. A. Bunnelle, *LC-GC* **9**(5) (1991) 332.
3. L. Schmidt and J. S. Fritz, *J. Chromatogr*. **640** (1993) 145.
4. P. J. Dumont and J. S. Fritz, *J. Chromatrogr. A* **691** (1995) 123.
5. L. Schmidt, J. J. Sun, J. S. Fritz, D. F. Hagen, C. G. Markell, and E. E. Wisted, *J. Chromatogr*. **641** (1993) 57.
6. D. L. Mayer and J. S. Fritz, *J. Chromatogr. A* **773** (1997) 189.
7. B. Bryan, *Today's Chemist* **2** (1994) 39.
8. D. D. Blevins and S. K. Schultheis, *LC-GC* **12** (1994) 12.
9. D. L. Ambrose and J. S. Fritz, *J. Chromatogr. B* **709** (1998) 89.
10. G. L. Lensmeyer, D. A. Wiebe, and T. C. Doran, *Ther. Drug Monit.* **14**(5) (1992) 408.
11. G. L. Lensmeyer, C. Onsager, I. H. Carlson, and D. A. Wiebe, *J. Chromatogr. A* **691** (1995) 259.
12. R. A. Grohs, F. V. Warren, Jr., and B. A. Bidlingmeyer, *Anal. Chem*. **63** (1991) 384.
13. G. R. Granneman and L. T. Sennello, *J. Chromatogr*. **229** (1982) 149.
14. A. Zlatkis, R. S. Brazell, and C. F. Poole, *Clin. Chem*. **27** (1981) 789.
15. K. E. Matsumoto, A. B. Robinson, L. Pauling, R. A. Flath, T. R. Mon, and R. Teranishi, in *Techniques of Combined GC/MS: Applications in Organic Analysis*, W. H. McFadden, ed., Wiley, New York, 1973, p. 375.
16. A. Zlatkis, K. Y. Lee, C. F. Poole, and G. Holzer, *J. Chromatogr*. **163** (1979) 125.
17. H. M. Liebich and J. Woll, *J. Chromatogr*. **142** (1977) 505.
18. T. A. Dirksen, S. M. Price, and S. J. St. Mary, *Am. Lab*. (Dec., 1993).
19. G. Font, J. Mañes, J. C. Moltó, and Y. Picó, *J. Chromatogr*. **642** (1995) 135.
20. J. J. Richard and G. A. Junk, *Mikrochim. Acta* **1** (1986) 387.
21. L. Schmidt, J. J. Sun, J. S. Fritz, D. F. Hagen, C. G. Markell, and E. E. Wisted, *J. Chromatogr*. **641** (1993) 57.
22. A. Kraut-Vass and J. Thoma, *J. Chromatogr*. **538** (1991) 233.
23. R. J. Bushway, H. L. Hurst, L. B. Perkins, L. Tian, C. Gulberteau Cabanillas, B. E. S. Young, B. S. Ferguson, and H. S. Jennings, *Bull. Environ. Contam. Toxicol*. **49** (1992).
24. J. F. Pankow, M. P. Ligocki, M. E. Rosen, L. M. Isabelle, and K. M. Hart, *Anal. Chem*. **60** (1988) 40.
25. Ch. W. Carter and I. H. Suffet, *Environ. Sci. Technol*. **16** (1982) 735.
26. P. F. Landrum, S. R. Nihart, B. J. Eadle, and W. S. Gardner, *Environ. Sci. Technol*. **18** (1984) 187.
27. C. T. Chiou, R. L. Malcolm, T. I. Brinton, and D. E. Kille, *Environ. Sci. Technol*. **20** (1986) 502.
28. J. H. Raymer and E. D. Pellizzari, *Anal. Chem*. **59** (1987) 1043.
29. J. H. Raymer, E. D. Pellizzari, and S. D. Cooper, *Anal. Chem*. **59** (1987) 2069.
30. L. J. Barnabas, J. R. Dean, S. M. Hitchen, and S. P. Owen, *Anal. Chim. Acta* **29** (1994) 261.

31. E. J. Poziomek, D. L. Eastwood, R. I. Lidberg, and G. Gibson, *Anal. Lett.* **24**(10) (1991) 1913.
32. C. Markell and E. Wisted, 3M Company, 1995.
33. J. S. Fritz, P. J. Dumont, and L. W. Schmidt, *J. Chromatogr. A* **691** (1995) 133.

# CHAPTER 7

# PRECONCENTRATION OF METAL IONS

## 7.1 INTRODUCTION

Cleanup of analytical samples prior to the determination of various metal ions has been a long-standing problem in chemical analysis. This sample pretreatment has two main goals. One is to separate the target metal ions from large amounts of salt, organic material, dirt, or other sample substances that would interfere in the analytical measurement step. A second goal is to preconcentrate the metal ions to be determined to a point where their analytical determination is facilitated.

The need for sample pretreatment is well illustrated by inductively coupled plasma–mass spectroscopy (ICP-MS), a widely used and relatively new technique for trace multielement and isotopic analysis of solutions (1–4). It is applicable to a wide range of samples, although highly saline samples can cause both spectral interferences and matrix effects (5,6). Changes in analyte count rates are observed with high levels of salts or heavy matrix ions. Orifice plugging is also a problem for samples of high solid content (≥0.5%) (7,8).

Spectral overlaps can also be a problem. The mass-spectral overlap of $MoO^+$ with all the useful isotopes of cadmium (9) and the overlap of $TiO^+$ with both copper isotopes and the most abundant zinc peak (10) are two examples. Finally, ionization interferences can result in significant quantitative errors in ICP-MS. Although almost any element is capable of inducing such an interference, uranium can be one of the worst (11,12).

Difficulties in connection with metal analysis by ICP-MS and other spectral methods are alleviated by using SPE to isolate the desired metal ions prior to analysis (13,14) or to remove a matrix element, such as uranium(VI), from the trace metals to be measured (13).

For many years liquid–liquid extraction has been a popular method for sample pretreatment. Extraction of metal ions from seawater, soil extracts, and a myriad of other solutions by liquid–liquid extraction of metal dithiocarbamate complexes has been a ubiquitous analytical procedure. But such procedures are labor-intensive and are subject to a number of difficulties. The latter includes the frequent formation of emulsions that are slow to break up, impurities introduced by liquid organic solvents, the need to concentrate the organic extract by evaporation, and air- and water-pollution problems stemming from the use of substantial quantities of organic solvents.

The use of solid-phase extraction is an attractive way to avoid many of the difficulties associated with liquid–liquid extraction. Uptake of metal ions by a SPE column or membrane tends to be more complete than liquid–liquid extraction because multiple equilibria occur in column methods. In some instances no organic solvents at all are used in SPE of metal ions. Several possibilities exist for SPE of metal ions:

1. Ion exchange, including uptake of metal cations or anions by simple ion exchange, uptake of ionic metal complexes on an ion-exchange resin, and selective elution of metal ions from an ion exchanger by inorganic or organic complexing reagents.
2. Addition of reagents to form an ion-association complex or a neutral metal–organic complex, followed by SPE of the complex by a nonionic solid extractant.
3. Retention of selected metal ions by use of a chelating resin.

Each of these techniques is discussed in this chapter.

## 7.2 ION-EXCHANGE METHODS

### 7.2.1 Introduction and Principles

There is a voluminous literature on the separation of metallic cations and anions by ion-exchange chromatography. The modern form, known as *ion chromatography*, is covered in books (15–18) as well as in numerous journal papers. The literature contains almost no publications for isolation of metal ions that are labeled specifically as *solid-phase extraction*. The practical question is how to use or convert ion-exchange chromatographic methods to rapid, quick ways for isolation of metal species in a manner consistent with modern solid-phase extraction.

Sometimes there is a very wide variation in the affinity of an ion exchanger for different ions. Extensive data for retention of metal cations using dilute solutions of perchloric acid (19), hydrochloric acid (20), and others have been published by Strelow and co-workers and more recently by Sevenich and Fritz (21). Thus the stronger retention of +2 metal cations and still stronger retention of +3 cations compared to +1, such as $Na^+$ and $K^+$, can be used to retain the ions of higher charge on a very short cation-exchange column. A certain concentration of a strong acid is added to the sample so that the $H^+$ level is sufficient to wash the $Na^+$ and $K^+$ through the column because of competition for the $-SO_3^-$ exchange sites on the column.

Kraus and Nelson found that a number of metal ions form anionic chloro complexes in aqueous hydrochloric acid solutions and are strongly taken up by anion-exchange resins (22). In most cases a plot of distribution ratio against the molar concentration of hydrochloric acid starts at an acid concentration characteristic of that metal ion and increases as the hydrochloric acid concentration becomes greater. A figure showing such plots for most of the metallic elements was published in 1956 (22) and has been widely reproduced in various scientific books (23). Elements that do not form chloro complexes are not taken up by the anion exchanger.

Separations are generally achieved by adding the sample to an anion-exchange column in rather concentrated hydrochloric acid and eluting the nonsorbed metal ions with eluant of the same HCl concentration. Then the sorbed metal ions are eluted one at a time by stepwise reduction of the HCl strength of the eluant. For example, in 10–12 M HCl, cobalt(II) is retained by an anion-exchange resin as $CoCl_4^{2-}$ and iron(III) as $FeCl_4^-$, but $Ni^{2+}$ forms no complex and is washed through the column. Then cobalt is quickly eluted with 4 M HCl. A step shift to pure water then elutes the iron.

The key to adapting ion-chromatographic separations to the selective retention of a selected metal ion or group of ions is to find conditions where a "go/no-go" situation exists. By this, we refer to conditions where the desired ions are strongly taken up by the ion exchanger while everything else passes through. In some cases a series of solutions can be used to desorb the retained ions one or two at a time. Several specific examples will be given in the following section.

### 7.2.2 Examples of Selective SPE of Metal Ions

***7.2.2.1 Use of Hydrogen Peroxide Complexes*** Addition of a dilute solution of hydrogen peroxide to acidic samples converts several metal ions into stable hydrogen peroxide complexes that are not retained by a cation exchanger. Thus, hydrogen peroxide complexes of molybdenum(VI), tungsten(VI), niobium(V), and tantalum(V) pass quantitatively through a short cation-exchange column and are separated from most other metal cations that are retained on the cation-exchange column (24). Strelow used hydrogen peroxide and dilute sulfuric acid to separate titanium(IV) from more than 20 cations by cation exchange (25). Fritz and Abbink (26) used a dilute solution of hydrogen peroxide to elute both vanadium(IV) and vandium(V) from a cation exchange column and thus separate it from a number of metal ions.

***7.2.2.2 Bromide Complexes*** Most +2, +3, and +4 metal ions are cations and are taken up by a strong-acid cation exchanger, even from solutions that are fairly acidic. However, a few metal ions are converted to neutral or anionic halide complexes at low concentrations of an added acid. Fritz and Garralda (27) used this scheme to separate mercury(II), bismuth(III), and cadmium(II) by elution from a cation-exchange column with 0.1–0.5 M hydrobromic acid. Lead(II) is also eluted as a bromide complex by 0.6 M HBr (28).

***7.2.2.3 Fluoride Complexes*** Several metal ions react with fluoride in aqueous solution to form complexes that are strongly taken up by an anion-exchange resin column (29–32). However, it is also possible to separate metal ions that form fluoride complexes from those that do not by selective elution form a short cation-exchange column. The following metal ions form fluoride complexes that are readily eluted from a cation exchanger by 0.1 M HF (33): $Al^{3+}$, Mo(VI), Nb(V), Sn(IV), Ta(V), Ti(IV), U(VI), W(VI), and Zr(IV). Some 23 metal ions are retained on the cation exchanger. Iron(III), gallium(III) and a few other metal ions are eluted by 1.0 M HF.

***7.2.2.4 Effect of Organic Solvents*** Metal cations form halide complexes much more readily in organic solvents than in water. It is apparently easier to replace cationic aquo complexes with anionic halide complexes when the solvent contains less water. For example, the pink cobalt(II) cation requires around 4 or 5 M aqueous hydrochloric acid to be converted to a blue cobalt(II) chloride anion. In predominantly acetone solution, an intense blue color is formed with cobalt(II) in very dilute hydrochloric acid. Thus, the scope of ion-exchange group separations is increased greatly by carrying out separations in a mixture of water and an organic solvent.

Fritz and Rettig (34) showed that zinc(II), iron(III), cobalt(II), copper(II), and magnesium(II) can be separated from each other on a short cation-exchange column with eluants containing a fixed, low concentration of HCl and increasing the water–acetone proportion from 40% to 95% acetone in steps.

Halide complexes are generally taken up strongly by anion-exchange resins. Korkisch, Fritz, Strelow, and others have published extensively on anion-exchange separations in partly nonaqueous solutions. A method worked out by Korkisch and Hazan (35) is particularly valuable as a group separation of metals that form chloride complexes from those that do not. The method uses an eluant consisting of 90–95% methanol–0.6 M hydrochloric acid and requires only a short anion-exchange column. Thus we have a go/no-go situation and excellent group separations are obtained.

## 7.3 EXTRACTION OF ION-ASSOCIATION COMPLEXES

The extraction of iron(III) from aqueous solution containing a high concentration of hydrochloric acid into diethyl ether is one of the oldest inorganic separations known. Over the years it has been demonstrated that diisopropyl ether or methylisobutyl ketone are much better solvents for this extraction. The species extracted is actually an ion pair: $H_3O^+/FeCl_4^-$. The organic solvent used must contain an oxygen atom that solvates the ion pair through interaction with iron in the ion pair.

A complex anion formed between a metal ion and a complexing inorganic anion such as chloride is known as an *ion-association complex.* Actually, a number of metal cations form ion-association complexes with chloride, bromide, iodide, fluoride, nitrate, sulfate, or thiocyanide. Many of these are extractable as ion pairs into a suitable organic solvent.

A technique known as *extraction chromatography* was developed and used extensively for inorganic separations, particularly during the 1960s and 1970s. Many of these systems were concerned with the use of ion-association systems. In extraction chromatography, a porous organic polymer typically is impregnated with an organic liquid, which serves as the extractive solvent, and packed into a short column. An appropriate acid (HCl, $HNO_3$, HF, etc.) is added to the aqueous sample to provide conditions suitable for formation of ion-association complexes. Passing the sample through the packed column gives intimate contact of the aqueous phase with the immobilized organic solvent. This results in extraction of the metal complexes as a sharp band on the column. After washing the nonextracted metal ions through the column with acid of the same concentration as the sample, conditions are adjusted so that the retained metal ions will be quickly eluted from the column. Often pure water will cause the ion-association complexes to break up and the retained metal ions to be eluted. In other cases a stepwise elution of various retained metal complexes may be possible.

### 7.3.1 Examples of Extraction Chromatography

Many of the methods of extraction chromatography are relatively old publications. Bibliographies of the earlier work have been published (36,37) and a book on extraction chromatography appeared in 1975 (38).

The applications described below are illustrative of the useful and selective extractions of metal ions that can be obtained by extraction chromatography. Use of more modern polymeric resins in a minicolumn should give even better results than reported in the original references. Thus, a porous, spherical polystyrene resin will be superior to nonspherical materials such as Teflon 6 and Haloport-F, which are no longer available. The smaller, more uniform particle size of modern resins will permit the use of smaller, shorter extraction columns and, consequently, smaller volumes of eluting solvent.

***7.3.1.1 SPE of Metal–Chloride Complexes*** Iron(III) can be separated from many elements by extraction from 6 to 8 M hydrochloric acid (39). The extraction is carried out by passing the aqueous hydrochloric acid mobile phase through a column packed with Haloport-F, a dispersion polymer of tetrafluoroethylene, to which a stationary phase of 2-octanone is sorbed. Fluoride, phosphate, or citrate causes no interference. Traces of iron(III) can be separated quantitatively from high concentrations of copper(II) or zinc(II), and traces of titanium(IV) can be separated from large amounts of iron(III).

In subsequent papers a granular polymeric material was impregnated with methylisobutylketone (MIBK) and used to extract molybdenum(VI) from 1 to 1 hydrochloric acid containing 1 M sulfuric acid (40). Iron(III) is also extracted. The retained molybdenum(VI) was eluted from the column with 1:1 hydrochloric acid, then the iron(III) was stripped from the column with dilute sulfuric acid. In a similar method, tin(IV) was extracted onto a column impregnated with MIBK from aqueous 8 M hydrochloric acid (41). Using this method tin(IV) was separated quantitatively from bismuth(III), cadmium(II), copper(II), lead(II), mercury(II), and zinc(II). For the separation of tin(IV) and molybdenum(VI), the aqueous eluting phase is 1 M hydrochloric acid-3 M sulfuric acid. A column impregnated with isopropyl ether quantitatively retains antimony(V) from 8 M hydrochloric acid and permits tin(IV) to be eluted rapidly.

***7.3.1.2 Extraction of Metal–Fluoride Complexes*** The separation of niobium(V) and tantalum(V) from other metal ions and from each other is one of the most difficult of all classic analytical procedures. However, studies on the extraction of fluoride complexes into MIBK revealed possibilities for column separations. A chromatographic technique was devised in which niobium(V), tantalum V, molybdenum(VI), and tungsten(VI) were extracted onto a column impregnated with MIBK from aqueous mixtures of hydrofluoric, hydrochloric, and sulfuric acids (42). Subsequent elution with eluants of different compositions allowed each of these elements to be eluted separately. As little as 1 part of each of these metals was separated successfully from $10^6$ to $10^7$ times as much as all the others.

***7.3.1.3 Separation of Uranium(VI)*** Uranium(VI) is extracted selectively from aqueous ~6 M nitric acid into MIBK or tributylphosphate. However, the liquid–liquid extraction is far from quantitative unless a high concentration of a salting-out reagent is also added to the aqueous phase. In a related method, uranium(VI) was separated quantitatively from other metals using a silica gel column treated with 6 M nitric acid. Uranium was eluted quickly and selectively from the column with MIBK (43). In this method the phases are reversed from the previous examples of extraction chromatography; the organic solvent is now the eluting phase.

The distribution ratio for extraction of uranium(VI) from 6 M aqueous nitric acid into an equal volume of MIBK is only 1.10 (43), a value that would give only 52% extraction. But in the silica gel column, which has several theoretical plates, the uranium(VI) is eluted quantitatively by a rather small volume of MIBK. It is interesting that thorium(IV), which normally is partially extracted by MIBK, is not partially eluted from the column. The answer seems to be that silica gel has ion-exchange properties that are sufficient to retain quadrivalent metals such as thorium(IV), titanium(IV), and zirconium(IV), even from strongly acidic solutions (44).

The silica gel method can be used to isolate uranium from numerous other metal ions, or it can be used to separate traces of other metal ions from large amounts of uranium. In the latter case the retained metal ions were eluted from the column with 6 M aqueous nitric acid or with aqueous sulfuric acid in some cases.

***7.3.1.4 SPE of Gold*** Gold(III) is known to be strongly extracted from aqueous HCl into MIBK and other solvents as the ion-association complex, $H_3O^+AuCl_4^-$ (45). A porous polyacrylate resin that contains an ester group, Rohm and Haas XAD-7 (now available as an Amberchrome resin), extracts gold without using any additional organic solvent (46). Gold(III) is also adsorbed by the resin from aqueous nitric acid solutions.

In the published procedure (46), gold(III) in a 1.0 M aqueous solution was passed through a short column packed with XAD-7 resin. The gold was adsorbed as a tight, yellow band. After a brief wash, the gold was stripped from the column by a small volume of acetone–1 M aqueous HCl (2.5:1). The amount of gold was determined spectrophotometrically at 340 nm.

Excellent quantitative recoveries of gold were obtained by this procedure, even when a 1.0-liter sample of $2.8 \times 10^{-6}$ M gold(III) (0.55 ppm) was used. The method is also extremely selective. No interference was encountered from any of the following metal ions that were present in a 36–1000:1 mol ratio to gold: Al(III), Bi(III), Ca(II), Cd(II), Ce(IV), Co(II), Cr(III), Cu(II), Fe(III), Hg(II), Mn(II), Ni(II), Pb(II), Pt(IV), Ru(IV), Sb(V), Sn(IV), U(VI), and Zn(II). It is significant that none of the platinum group elements tested were extracted, because these elements all form stable, anionic chloro complexes.

## 7.4 EXTRACTION OF METAL–ORGANIC COMPLEXES

### 7.4.1 Principles

A very large number of organic reagents forms complexes with various metal ions. These reagents have many applications in analytical chemistry including use as color-forming reagents for spectrophotometry, precipitating reagents, and separation of isolation of metal ions by liquid–liquid extraction of their complexes. A book by Sandell and Onishi (47) discusses the chemistry and analytical applications of many organic complexing reagents in some detail.

8-Hydroxyquinoline, commonly known as "Oxine," forms complexes with an unusually large number of metal ions, and its applications in chemical analysis have been studied exhaustively. Oxine forms chelates, binding metal ions through the ring nitrogen and phenolate oxygen. A proton is lost from the phenol group as a result of chelation. Addition of excess Oxine to a metal ion in solution almost always produces a neutral metal–Oxine chelate.

$$Zn^{2+} + 2H\ Oxine \rightarrow Zn(Oxine)_2 + 2H^+$$

$$Al^{3+} + 3H\ Oxine \rightarrow Al(Oxine)_3 + 3H^+$$

Sandell and Onishi (47) give the pH range for precipitation, and in many cases the conditions for solvent extraction, of some 60 metal–Oxine complexes.

For any given chelating reagent, selectivity for desired metal ions can often be adjusted by control of pH and use of auxiliary complexing reagents. As pH is made more acidic, only those metal ions that form more stable complexes will still react. Addition of EDTA as an auxiliary reagent will form nonextractable complexes for most metal ions and prevent their reaction with Oxine. However, uranium(VI) is only weakly complexed by EDTA but forms an extractable Oxine complex. This provides a very selective way to extract uranium.

Solid-phase extraction is again preferred to liquid–liquid extraction for isolation of metal ions as metal–organic complexes. Two SPE modes are feasible. One way is first to adsorb the chelating reagent onto the solid-phase extractant particles. This can be done either before or after the particles are packed into a minicolumn. Then the aqueous sample is passed through the minicolumn. This causes the metal ions in solution to come into intimate contact with a comparatively high concentration of the chelating reagent on the resin surfaces. In this mode the contact time is relatively short, so the rate of chelate formation must be rapid.

A second SPE mode is to add a chelating reagent to the aqueous sample and adjust the pH to a suitable value. At least a two- to threefold excess of reagent must be used to ensure complete chelation of the desired metal ions. In sample solutions containing very dilute concentrations of metal ions to be extracted a much larger excess of reagent may be needed to avoid partial dissociation of the metal chelates. After complexation in solution, the sample is simply passed through the SPE column to extract the metal chelates. In practice, this method has often been found to give better results than the first operational mode. Addition of the reagent to the sample solution allows adequate time for metal–organic complexes to form prior to the SPE step.

There are two major possibilities for desorption of extracted complexes. One is simply to wash the SPE column with a small volume of an organic solvent, such as methanol or acetone. This will usually dissolve the metal chelates and cause them to elute quickly and completely. Another possibility is to wash the SPE column with an aqueous solution that is sufficiently acidic to break up the metal–organic complex. The free metal ions will no longer be retained by the resin. A combination of these two approaches—a solution of a strong mineral acid in an organic solvent—will simultaneously break up the complex and elute both the metal ions and the bulk of the complexing reagent.

### 7.4.2 Selection and Design of Chelating Reagent

For best results, the chelating reagent should fulfill several criteria.

1. The chelating group in the reagent should be selective for the target metal ions. It is preferred that common metal ions such as $Na^+$, $Mg^{2+}$, and $Ca^{2+}$ not be taken up.
2. The metal chelates should be stable; that is, the chelates should not break up even when the concentration of metal ions in the sample is very low.
3. The rate of formation of the metal–organic chelates must be rapid when the chelating reagent is immobilized on the SPE column. Fast kinetics are not so important when the reagent is added to the sample prior to the SPE step.
4. The chelating reagent should be soluble in water or in water containing some organic solvent. The metal chelates should also be soluble in the sample solution because precipitates are extracted more slowly.
5. Although the reagent and the chelate should both be soluble, the metal chelates must be strongly extracted by the solid-phase particles.

It is often better to do some simple "molecular engineering" on the chelation reagent in order to obtain desirable solubility characteristics.

**TABLE 7.1 Some Chelating Reagents for SPE of Metal Ions**

| Class | Examples | Compound |
|---|---|---|
| Dithiocarbamate | N—C(=S)—$S^-NH_4^+$ | I |
| Dithiocarbamate | $(HOCH_2CH_2)_2N$—C(=S)—$S^-Na^+$ | II |
| Hydroxamic acid | C(=O)—N(OH)—$CH_3$ | III |
| Pyrazolone | $H_3C$—C(=O)—CH; C(=O); N—$CH_3$; $H_3C$—C=N | IV |
| Pyridylazoresorcinol | N; —N=N—; OH; HO | V |

Table 7.1 lists several classes of metal chelates that have been used in SPE, together with some specific reagents.

### 7.4.3 Examples of SPE with Metal Chelates

***7.4.3.1 Ammonium Pyrolidine Dithiocarbamate*** Compound I in Table 7.1 was used in an on-line preconcentration system for electrothermal atomic absorption (graphite furnace) (48). A microscale SPE 2.1-mm-ID column was filled with a bed of extractant particles 3 to 4 mm high. Silica C18, Rohm and Haas XAD-2 and XAD-7, all worked well. A mixing "T" just before the microcolumn mixed sample and reagent together; the metal complexes were adsorbed by the particles in the micro column. Acetonitrile was then used to desorb the metal complexes.

This method was used to determine Cd, Co, Cu, Fe, Ni, and Pb in Antarctic seawater. The blank level was very low and the detection limits ranged from 0.4 ng/liter (Cd) to 25 ng/liter (Fe).

***7.4.3.2 Sodium Bis(2-hydroxyethyl)dithiocarbamate*** Compound II in Table 7.1 was used to form neutral metal complexes with metal ions of the hydrogen sulfide group (49). The hydroxy groups in the molecule cause the metal ion complexes to be water-soluble at low concentrations. The complexed metal ions can be concentrated from a very dilute solution by sorption on a small column of XAD-4 resin. Additional selectivity was achieved by pH adjustment or by the use of masking agents. The sorbed metal complexes were eluted with acidic ethanol for subsequent analytical measurement.

The effect of pH on the extraction of metal dithiocarbamates is shown in Table 7.2. All the metal ions studied were taken up quantitatively between pH 6.0 and 7.0 except for molybdenum(VI), which was extracted quantitatively at pH 3.0 or 4.0. EDTA, as a masking agent, prevented extraction of iron(III), nickel(II), or zinc(II) while allowing 100% extraction of silver(I), bismuth(III), copper(II), and mercury(II). Cyanide at pH 9.0 masked cobalt(II), copper(II), nickel(II), and zinc(II), but SPE of bismuth, cadmium(II), and lead(II) was still complete.

A study of breakthrough capacity showed that sorbents with higher surface area retained the metal dithiocarbamates more strongly (49). The polystyrene resin XAD-4 with surface area of ~725 $m^2/g$ was by far the

**TABLE 7.2 Percentage Recovery of Metal Dithiocarbamates as a Function of pH**

| | Solution pH | | | | | | | |
|---|---|---|---|---|---|---|---|---|
| Element | 3.0 | 4.0 | 5.0 | 6.0 | 7.0 | 8.0 | 9.0 | 10.0 |
| Ag(I) | 95 | 100 | 100 | 100 | 100 | 100 | 100 | 100 |
| Cd(II) | 19 | 100 | 100 | 100 | 100 | 100 | 100 | 100 |
| Fe(III) | 0 | 52 | 100 | 100 | 100 | 100 | 84 | 31 |
| Mo(VI) | 100 | 100 | 73 | 37 | 0 | 0 | 0 | 0 |
| Ni(II) | 74 | 100 | 100 | 100 | 100 | 100 | 100 | 100 |
| U(VI) | 0 | 0 | 33 | 100 | 100 | 33 | 16 | 13 |
| V(V) | 100 | 100 | 100 | 100 | 100 | 82 | 53 | 14 |
| Zn(II) | 0 | 1 | 100 | 100 | 100 | 100 | 100 | 100 |

*Source:* From Reference 49 with permission.

best. XAD-7, a polyacrylate resin with area ~450 $m^2/g$, was the best of the polyacrylate sorbents studied.

More recent work (50) has demonstrated significant improvements. In the original work, ground-up resins were used, sieved to a particle size of 74–105 μm, in a 175 × 9-mm glass column (49). Later, a 5-mm-ID mini column was used, packed with 10 μm PS-DVB resin to a height of only 3 mm (50).

A 10-ml sample volume containing 0.3 ppm of the metal at pH 4 was passed through the column. Subsequent desorption of the metal complexes was readily obtained by 1 ml of 1 M nitric acid in ethanol.

Solid NaHEDC requires rather careful preparation, purification, and storage to prevent oxidation. In the latter work, these problems were virtually eliminated by preparing the reagent in solution by a procedure similar to that originally proposed by Fritz and Sutton (51). Diethanolamine (1.0 ml) and carbon disulfide (30 μl) are added to 5–10 ml of methanol. The container is stoppered and allowed to stand for a few minutes to allow the reaction to be completed. The solution is then diluted to 25 ml with methanol. This solution showed almost no change in DTC concentration for at least 16 days and remained clear and usable for at least 2 months.

Contamination of complexing reagents by metal ions is often a problem. Most organic chemicals and solvents contain significant amounts of various inorganic impurities. Metal ions that may be adsorbed on the container walls may enter the solution by the complexing action of the reagent. The reagent blank can be reduced to very low levels simply by passing the DTC solution through a resin column to remove the complexed impurities. Typical values of the reagent blank (ng/ml) are Cd(II) < 0.02; Cu(II), 0.6; Fe(III), 1.0; Hg < 0.02; Pb(II) < 0.2; and Zn, 0.6 (50).

***7.4.3.3 Hydroxamic Acids*** Hydroxamic acids, and *N*-phenylbenzohydroxamic acid (PBHA) in particular, have found extensive use as precipitants and as complexing agents for isolation of metal ions by solvent extraction (52–54). Hydroxamic acid complexes of metals in the +6, +5, +4, and some +3 oxidation states are unusually stable and can be formed at very acidic pH values. However, the metal complexes of PBHA and several other hydroxamic acids are insoluble in aqueous solution. For SPE, a reagent is needed such that the metal chelates have a reasonable solubility in water but are strongly sorbed by an appropriate resin.

After several hydroxamic acid reagents had been synthesized and tested, *N*-methylfurohydroxamic acid (MFHA) (compound III in Table 7.1) was selected by Albiaty and Fritz for SPE (55). It is easily prepared and its metal complexes are soluble in water at low concentrations but can be sorbed by XAD-4 resin. A procedure that is extremely effective for concentrating metal ions prior to their measurement was developed.

Considerable selectivity was possible in the isolation of metal ions by manipulation of sample pH. The following ions were extracted quantitatively as MFHA complexes from 1 liter of $10^{-6}$ M metal ion concentrations at pH 0–0.5: Fe(III), Mo(VI), Sc(III), Sn(IV), Ti(IV), W(VI), and Zr(IV). At pH 5.0 the following additional elements were extracted: Al(III), Bi(III), Cr(III), Cu(II), Th(IV), U(VI), and V(IV). Several additional elements were extracted at pH 9.0.

The very strong chelating ability of MFHA was demonstrated by experiments in which low ppm concentrations of several ions were isolated by SPE of their hydroxamic acid complexes from 5% and 10% aqueous solutions of calcium nitrate, 10% aluminum chloride, and 5% sodium acetate. Results are summarized in Table 7.3.

MFHA has also been used in a SPE method to isolate molybdenum(VI) and titanium(IV) and avoid overlapping mass spectral peaks with copper, cadmium, or zinc (13).

*7.4.3.4 Extraction of Metal–Pyrazolone Complexes* Since their introduction as analytical reagents by Jensen in 1959 (56–58), the 4-acyl-2-pyrazolin-5-one class of compounds has found increasing use for the liquid–liquid extraction of metal ions from aqueous solution. The most popular reagent of this class is 1-phenyl-3-methyl-4-benzoyl-2-pyrazolin-

**TABLE 7.3 Removal of $10^{-6}$ M Metal Ion Impurity Added to a Concentrated Salt Solution**

| Salt | Concentration, % | Volume, ml | Metal Added | pH | Recovery, % |
|---|---|---|---|---|---|
| $Ca(NO_3)_2 \bullet 4H_2O$ | 5 | 10.0 | Fe(III) | 5.0 | 100 |
| $Ca(NO_3)_2 \bullet 4H_2O$ | 10 | 100.0 | Fe(III) | 5.0 | 95 |
| $Ca(NO_3)_2 \bullet 4H_2O$ | 5 | 10.0 | Zn(II) | 9.0 | 100 |
| $Ca(NO_3)_2 \bullet 4H_2O$ | 5 | 10.0 | Mn(II) | 9.0 | 100 |
| $Ca(NO_3)_2 \bullet 4H_2O$ | 5 | 10.0 | Co(II) | 9.0 | 90 |
| $Ca(NO_3)_2 \bullet 4H_2O$ | 5 | 10.0 | Ni(II) | 9.0 | 40 |
| $Ca(NO_3)_2 \bullet 4H_2O$ | 1 | 10.0 | Ni(II) | 9.0 | 90 |
| $Al(Cl)_3 \bullet 6H_2O$ | 10 | 10.0 | Zr(IV) | 0.5 | 100 |
| $CH_2COONa \bullet 3H_2O$ | 5 | 10.0 | Cu(II) | 5.0 | 100.5 |

*Source:* From Reference 55 with permission.

5-one (PMBP). The uses of PMBP as an extractant have been reviewed by Minczaewski et al. (59) and by Zolotov and Kuzmin (60). For isolation of metal ions by solid-phase extraction, a reagent is needed where the metal chelates are more soluble than with PMBP. King and Fritz (61) described the synthesis and analytical applications of a more soluble reagent, 1,3-dimethyl-4-acetyl-2-pyrazolone (DMAP) (compound IV in Table 7.1).

A small column packed with XAD-4 resin was used for SPE. A 10:1 excess of DMAP was added to the sample together with an appropriate buffer and tartrate to form a complex with metal ions and avoid any precipitation. Essentially 100% recoveries were obtained for Th(IV) and Zr(IV) in the pH range 2–6 and for U(VI) in the range 3–6. High recoveries were also obtained for Co(II), Cu(II), Fe(III), La(III), and Ni(II) at pH 5–6. By operating at pH 5.0 with 0.001 DMAP and 0.001 M EDTA, it was possible to isolate 10 ppm uranium(VI) from a 1.0 liter sample containing 100–1000 times as much (molar concentrations) of the following elements: Co(II), Cu(II), Fe(III), La(III), Mn(II), Ni(II), Th(IV), Zn(II), and Zr(IV).

***7.4.3.5 Other Organic Complexing Reagents*** The examples given above certainly do not exhaust the possibilities of using an organic chelating reagent to extract various metal ions selectively by SPE. Roston (62) used PAR (compound V in Table 7.1) for the liquid chromatographic separation of several metal ions. However, a preliminary liquid extraction with chloroform was used to isolate the PAR complexes from the aqueous sample prior to the chromatographic separation.

8-Hydroxyquinoline forms extractable complexes with many metal ions. This reagent has been used in conjunction with a small SPE column to isolate various metal ions (63).

## 7.5 CHELATING RESINS

### 7.5.1 Introduction

In principle, the ideal way to isolate and preconcentrate metal ions is to pass the sample through a small column or membrane containing a chelating resin. The desired metals would be plucked out of solution by the chelating groups of the solid particles. Passing a highly acidic solution through the resin to break up the metal chelates on the solid resin should result in rapid elution of the metal ions. Sometimes SPE with chelating resins does, indeed, work this well. But the relatively large particle size of chelating polymers and slow mass transport within the polymer beads has frequently led to procedures that often are rather awkward and slow.

There certainly is a huge number of chelating resins for removal of various metal ions from solution. Examples of chelating groups and their application for preconcentration of inorganic elements were reviewed critically by Myasoedova and Savvin (64). Chow and co-workers reviewed the use of chelating polymers and related supports for the preconcentration of trace metals from natural waters (65). Schwochau documented the use of chelating solids for the extraction of metals from seawater (66). An entire section was devoted to the extraction of uranium from seawater.

In the following sections a few specific examples are given to illustrate the very extensive possibilities of SPE with chelating resins.

### 7.5.2 Iminodiacetic Acid (IDA) Resins

A Dow A-1 chelating resin, also known as *Chelex 100*, is perhaps the oldest and most popular of the chelating resins used in chemical analysis. It contains the following functional group:

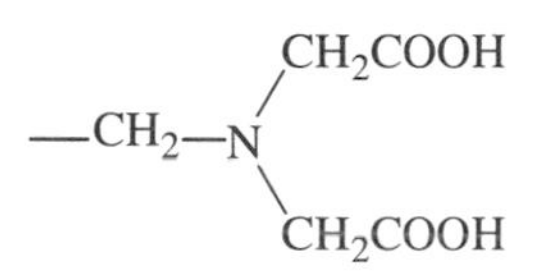


It combines with many metal ions through the nitrogen and the two carboxylate groups. Although Chelex IDA resins have been used successfully for isolation of trace metal ions from seawater (67) and other applications, the properties of this resin are far from ideal. The capacity for complexing metal ions is high, but the resin particles are rather large and the low degree of polymeric crosslinking leads to excessive swelling or shrinking of the resin bed in a column when large pH changes occur. An ideal chelating resin for SPE might have the following properties: a high degree of crosslinking for good physical stability, and small spherical particles (perhaps ~10 μm in diameter), porous with a high surface area for fast kinetics, and with the chelating groups on or near the surface to avoid steric hinderance in metal ion chelation.

A high concentration of IDA groups on the chelating resin results in more complete complexation of metal ions from solution. A high concentration of complexing groups may also cause the resin to retain metal ions from a more acidic sample. However, stronger complexation of metal ions means that a more concentrated acid solution must be used to break up the complex and thereby desorb metal ions from the resin. The presence of excess acid may complicate the determination of sample ions by ion chromatography or capillary electrophoresis.

An additional complication is that two kinds of metal ion uptake can occur with IDA resins. The desired kind of uptake involves chelation of metal ions with the nitrogen and carboxyl groups of the IDA as ligands. The other type is simple ion exchange of cations that are electrostatically attracted to the negatively charged carboxylate groups. This simple ion exchange can take up a significant amount of $Na^+$ or other unwanted cations, particularly if the resin contains a high concentration of IDA groups.

On balance, the best choice of an IDA resin might be one with a moderately low capacity (e.g., ~0.5 meq/g) for retaining metal ions by chelation. A resin particle size of ~10 μm, instead of the 40–50 μm size generally used in solid-phase extraction (SPE) cartridges, permits effective concentration of metal ions with a resin bed of only a few millimeters.

Gennaro et al. (68) describe a chemical method for introducing IDA groups into cellulose filters. The resulting filters were then used for SPE of several metal cations.

### 7.5.3 Hydroxamic Acid Resins

Phillips and Fritz described synthetic methods for attaching the following hydroxamic acid groups to the benzene rings in XAD-4 PS-DVB resin particles (69,70):

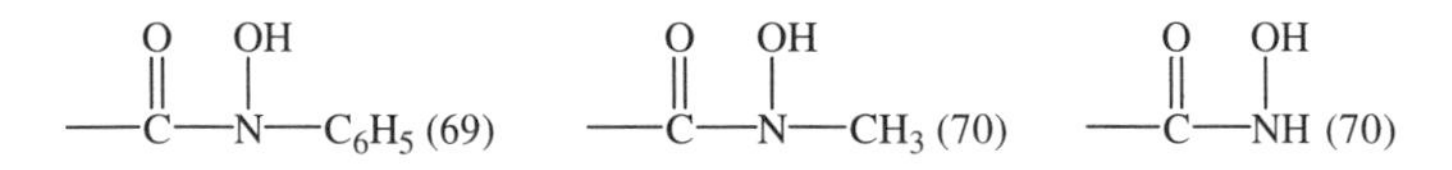


Some of the complexes are so stable that quantitative extraction of several metal ions was possible from very acidic sample solutions. The following metal ions were quantitatively taken up by short resin columns from pH 4 solutions: Cu(II), Al(III), Fe(III), and U(VI). Thorium(II) and Ti(IV) were quantitatively extracted from 0.1 M HCl; Mo(VI), W(VI), and Zr(IV) were completely taken up from 2 M HCl. Several of these metal ions required the use of a complexing eluant for desorption such as 0.1 M oxalic acid or 0.3 M HF.

Vernon and Eccles published a number of papers dealing with synthesis, properties, and applications of hydroxamic acid resins (71,72). Poly(amidoxime)/hydroxamic resins and fibers have been used for the extraction of uranium from seawater (73).

### 7.5.4 Thioglycolate Resins

Resins containing a thiol group (–SH) complex $Ag^+$, $Hg^{2+}$, $Pb^{2+}$ and other metal ions having an affinity for sulfur. Examples of this type include resins with a thioglycolate group ($-OCOCH_2SH$) attached to the benzene ring of XAD-4 polymer beads through a hexylcarboxylate bridge (74) or simply through a methylene bridge (75). Deretani and Sebille (76) described the synthesis and analytical applications of an acrylate resin containing a mercaptoacetamide chelating group ($-NHCOCH_2SH$).

Uptake of the following metal ions by resins of this type has been reported: $Ag^+$, Au(III), $Bi^{3+}$, $Cd^{2+}$, $Cu^{2+}$, $Fe^{3+}$, $Hg^{2+}$, $Pb^{2+}$, Sb(III), Sn(IV), $UO_2^{2+}$, and $Zn^{2+}$. Most of these metals are extracted from pH ~5, but $Ag^+$ and $Hg^{2+}$ are retained from very acidic solutions (~ pH 0–1).

### 7.5.5 Cellulose Phosphate Materials

Ion exchangers containing the phosphonate group $[-PO(OH)_2]$ or phosphate group $[-OPO(OH)_2]$ form very strong complexes with certain metal ions.

One of the best and most convenient materials of this class is Whatman cellulose phosphate P-11, which is commercially available. This has been used in column separations of uranium(VI) from other metal ions (77,78) and in the recovery of thorium(IV) from monazite ores (79). These metal ions are complexed by cellulose phosphate even from highly acidic solutions. Uranium(VI) is quickly and easily eluted from a short P-11 column by aqueous sodium carbonate, which forms a soluble carbonato complex with uranium (80).

Rare-earth metal ions are strongly retained by P-11 cellulose phosphate and can be separated from $Sr^{2+}$ from 0.5 M HCl (81). The alkaline-earth metal ions are also extracted from a somewhat less acidic solution. For example, $Sr^{2+}$ is separated from $Cs^+$ from 0.1 M HCl solutions (81).

## 7.6 SOME CONCLUSIONS

A major goal of this chapter has been to give the reader some appreciation of the rich literature that exists in the isolation and chromatographic separation of various metal ions. While some of the methods described are very usable as originally published, others will require some modification to achieve the rapid, efficient, and convenient solid-phase extractions needed in contemporary chemical analysis.

These modifications should be accomplished very easily in many cases. Thus, extraction of metal ion-association complexes or metal–organic chelates can be made more efficient merely by using resin-loaded membranes or minicolumns packed with a smaller bed of modern extractant particles. Synthesis of the water-soluble chelating regents is often simple, although it would be helpful if more of these were available commercially.

The use of chelating resins for SPE of metal ions presents more of a problem because scientists seldom want to take the time and trouble to prepare new materials. Actually, the synthesis of the noncommercial resins described in Section 7.5 is relatively simple. Hydroxamic acid and thiol resins can usually be made by a two-step organic reaction. The particle size of some older commercial chelating resins is too large and the general efficiency is too low for modern SPE. However, a few chelating resins with better performance are beginning to appear on the market.

## REFERENCES


1. R. S. Houk, V. A. Fassel, G. D. Flesch, H. J. Svec, A. L. Gray, and C. E. Taylor, *Anal. Chem.* **52** (1980) 2283–2289.
2. R. S. Houk and J. J. Thompson, *Mass Spectrom. Rev.* **7** (1988) 425–462.
3. R. S. Houk, *Anal. Chem.* **58** (1986) 97A–105A.
4. D. J. Douglas and R. S. Houk, *Prog. Anal. Atom. Spectrosc.* **8** (1985) 1–18.
5. S. H. Tan and G. Horlick, *Appl. Spectrosc.* **40** (1986) 445–460.
6. S. H. Tan and G. Horlick, *J. Anal. Atom. Spectrom.* **2** (1987) 745–763.
7. J. A. Olivares and R. S. Houk, *Anal. Chem.* **58** (1986) 20–25.
8. D. J. Douglas and L. A. Kerr, *J. Anal. Atom. Spectrom.* **3** (1988) 749–752.
9. C. W. McLeod, A. R. Date, and Y. Y. Cheung, *Spectrochim. Acta, Part B* **41** (1986) 169.
10. J. W. McLaren, D. Beauchemin, and S. S. Berman, *J. Anal. Atom. Spectrom.* **2** (1987) 277.
11. J. J. Thompson and R. S. Houk, *Appl. Spectrosc.* **41** (1987) 801.
12. D. Beauchemin, J. W. McLaren, and S. S. Berman, *Spectrochim. Acta, Part B* **42** (1987) 467.
13. S. J. Jiang, M. D. Palmieri, J. S. Fritz, and R. S. Houk, *Anal. Chim. Acta* **200** (1987) 559.
14. M. R. Plantz, J. S. Fritz, F. G. Smith, and R. S. Houk, *Anal. Chem.* **61** (1989) 149.
15. J. S. Fritz, D. T. Gjerde, and C. Pohlandt, *Ion Chromatography*, 1st ed., Hüthig, Heidelberg, 1982.
16. D. T. Gjerde and J. S. Fritz, *Ion Chromatography*, 2nd ed., Hüthig, Heidelberg, 1987.

17. F. C. Smith, Jr. and R. C. Chang, *The Practice of Ion Chromatography*, Wiley-Interscience, New York, 1983.
18. P. R. Haddad and P. E. Jackson, *Ion Chromatography, Principles and Applications*, Elsevier, Amsterdam, 1990.
19. F. W. E. Strelow and H. Sondorp, *Talanta* **18** (1972) 1113.
20. F. W. E. Strelow, *Anal. Chem.* **32** (1960) 1185.
21. G. J. Sevenich and J. S. Fritz, *J. Chromatogr.* **371** (1986) 361.
22. K. A. Kraus and F. Nelson, *Proc. 1st U.S. Int. Conf. Peaceful Uses Atom. Energy* **7** (1956) 113.
23. J. S. Fritz and G. H. Schenk, *Quantitative Analytical Chemistry*, 4th ed., Allyn and Bacon, Boston, 1979, p. 422.
24. J. S. Fritz and L. H. Dahmer, *Anal. Chem.* **37** (1965) 1272.
25. F. W. E. Strelow, *Anal. Chem.* **35** (1963) 1279.
26. J. S. Fritz and J. E. Abbink, *Anal. Chem.* **34** (1962) 1080.
27. J. S. Fritz and B. B. Garralda, *Anal. Chem.* **34** (1962) 102.
28. J. S. Fritz and R. C. Greene, *Anal. Chem.* **35** (1963) 811.
29. K. A. Kraus and F. Nelson, *ASTM Spec. Tech. Publ.* 195, 1958, p. 27.
30. F. Nelson, R. M. Rush, and K. A. Kraus, *J. Am. Chem. Soc.* **82** (1960) 339.
31. U. Schindewolf and J. W. Irvine, Jr., *Anal. Chem.* **30** (1958) 906.
32. L. Wish, *Anal. Chem.* **31** (1959) 326.
33. J. S. Fritz, B. B. Garralda, and S. K. Karraker, *Anal. Chem.* **33** (1961) 882.
34. J. S. Fritz and T. A. Rettig, *Anal. Chem.* **34** (1962) 1562.
35. J. Korkisch and I. Hazan, *Talanta* **11** (1964) 1157.
36. C. E. Hedrick and J. S. Fritz, *Bibliography of Reversed-Phase Partition Chromatography*, Iowa State Univ. Document IS-950, June 1, 1964.
37. H. Eschrich and W. Drent, *Bibliography on Application of Reversed-Phase Partition Chromatography to Inorganic Chemistry and Analysis*. Eurochemic, Mol, Belgium, Document ETR 211, Nov. 1967.
38. T. Braun and G. Ghersini, *Extraction Chromatography*, Elsevier, Amsterdam, 1975.
39. J. S. Fritz and C. E. Hedrick, *Anal. Chem.* **34** (1962) 1411.
40. J. S. Fritz and C. E. Hedrick, *Anal. Chem.* **36** (1964) 1324.
41. J. S. Fritz and G. L. Latwesen, *Talanta* **14** (1967) 529.
42. J. S. Fritz and L. H. Dahmer, *Anal. Chem.* **40** (1968) 20.
43. J. S. Fritz and D. H. Schmitt, *Talanta* **13** (1966) 123.
44. S. Ahrland, I. Grenthe, and B. Noren, *Acta Chem. Scand.* **14** (1960) 1059, 1077.
45. F.W.E. Strelow, E. C. Fest, P. M. Mathews, C. J. C. Bothma, and C. R. Van Zyl, *Anal. Chem.* **38** (1966) 115.
46. J. S. Fritz and W. G. Millen, *Talanta* **18** (1971) 323.
47. E. B. Sandell and H. Onishi, *Photometric Determination of Traces of Metals*, Wiley, New York, 1978.
48. V. Porta, O. Abbolino, E. Mentasti, and C. Sarzanini, *J. Anal. Atom. Spectrom.* **6** (1991) 119.

49. J. N. King and J. S. Fritz, *Anal. Chem.* **57** (1985) 1016.
50. J. S. Fritz, R. C. Freeze, M. J. Thornton, and D. T. Gjerde, *J. Chromatogr. A* **739** (1996) 57.
51. J. S. Fritz and S. A. Sutton, *Anal. Chem.* **28** (1956) 1300.
52. A. D. Shendrikar, *Talanta* **16** (1969) 51.
53. G. Grossi and F. Baroncelli, *2nd Proc. Int. Solvent. Extr. Conf.*, 1977, p. 640.
54. H. Forster and K. Schwabe, *Anal. Chim. Acta* **45** (1969) 511.
55. I. A. Albiaty and J. S. Fritz, *Anal. Chim. Acta* **146** (1983) 191.
56. B. S. Jensen, *Acta Chem. Scand.* **13** (1959) 1347.
57. B. S. Jensen, *Acta Chem. Scand.* **13** (1959) 1668.
58. B. S. Jensen, *Acta Chem. Scand.* **13** (1959) 1890.
59. J. Minczaewski, J. Chwastowska, and R. Dybczynski, *Separation and Preconcentration Methods in Inorganic Analysis*, Halsted Press, New York, 1982, pp. 193–195.
60. Y. A. Zolotov and N. M. Kuzmin, *Metal Extraction with Acylpyrazalones*, Izdat Nauka, Moscow, 1977.
61. J. N. King and J. S. Fritz, *Anal. Chim. Acta* **207** (1988) 137.
62. D. A. Roston, *Anal. Chem.* **56** (1984) 241.
63. R. E. Sturgeon, S. S. Berman, and S. N. Willie, *Talanta* **29** (1982) 167.
64. G. V. Myasoedova and S. B. Savvin, *Crit. Rev. Anal. Chem.* **17** (1986) 1.
65. C. Kantipuly, S. Katragadda, A. Chow, and H. D. Gesser, *Talanta* **37** (1990) 491.
66. K. Schwochau, *Top. Curr. Chem.* **124** (1984) 91.
67. R. Boniforti, R. Ferraroli, P. Frigieri, D. Heltai, and G. Queirazza, *Anal. Chim. Acta* **162** (1984) 33.
68. M. C. Gennaro, C. Baiocchi, E. Campi, E. Mentasi, and R. Aruga, *Anal. Chim. Acta* **151** (1983) 339.
69. R. J. Phillips and J. S. Fritz, *Anal. Chim. Acta* **121** (1980) 225.
70. R. J. Phillips and J. S. Fritz, *Anal. Chim. Acta* **139** (1982) 237.
71. F. Vernon and H. Eccles, *Anal. Chim. Acta* **77** (1975) 145; **79** (1975) 229.
72. F. Vernon and H. Eccles, *Anal. Chim. Acta* **82** (1976) 369; **83** (1976) 187.
73. F. Vernon and T. Shah, *React. Polym.* **1** (1983) 301.
74. E. M. Moyers and J. S. Fritz, *Anal. Chem.* **48** (1976) 1117.
75. R. J. Phillips and J. S. Fritz, *Anal. Chem.* **50** (1978) 1504.
76. A. Deretani and B. Sebille, *Anal. Chem.* **53** (1981) 1742.
77. T. Bruce and R. W. Ashley, *Analyst* **92** (1967) 137.
78. G. C. Goode and M. C. Campbell, *Anal. Chim. Acta* **27** (1962) 422.
79. A. J. Head, N. F. Kember, R. P. Miller, and R. A. Wells, *J. Appl. Chem.* **9** (1959) 599.
80. R. C. Freeze and J. S. Fritz, unpublished data, 1996.
81. D. H. Schmitt and J. S. Fritz, *Talanta* **15** (1968) 515.

# CHAPTER 8

# MICROSCALE AND SEMIMICROSCALE TECHNIQUES

Solid-phase extraction with packed minicolumns or membranes certainly has many attractive features compared with traditional solvent extraction. However, like any technique, SPE has its limitations. Finely divided solids, oily materials, or large biomolecules in the sample matrix can clog the pores of the solid sorbent. Even when SPE is performed with a column of very small dimensions, the organic solvent used to elute the analytes is apt to be at least 100 μl. Unless the volume of this eluate is reduced by careful (and tedious) evaporation, only a small fraction of the sample will actually be used for subsequent analysis by GC or HPLC. Two techniques that address this problem are discussed in this chapter. One is a technique known as *solid-phase micro extraction.* This is a truly micro technique in which analytes are adsorbed onto a fiber coated with a thin polymeric coating. The other method is a semimicro adaption of the conventional SPE in which a tiny membrane is held in the hub of a syringe needle.

## 8.1 SOLID-PHASE MICROEXTRACTION (SPME)

### 8.1.1 Introduction

SPME is a very simple microscale technique introduced by Belardi and Pawliszyn in the late 1980s (1). It has the advantage that *all* the analytes collected on the solid phase can be injected into a gas chromatograph for further analysis. This is accomplished by using a very small amount of extractive solid phase coated onto a silica fiber. Following the extractive step, this fiber can be introduced into the heated injection port of a gas chromatograph and the analytes thermally desorbed. An excellent review of the principles and applications of this technique has been published (2).

The device used for SPME is basically a modified syringe (see Fig. 8.1). A fused-silica fiber approximately 1 cm long is coated with a stationary phase of polysiloxane or some other polymer. The fiber is glued using high-temperature epoxy to a small stainless steel tube that runs through the syringe needle. The fragile fiber is initially withdrawn into the steel syringe needle, which protects the fiber as the septum of the sample container is pierced. After the sample septum is pierced, the coated fiber is extended into the sample solution for a set time (typically 2–15 min for liquid samples) where the analytes are adsorbed by the fiber coating. The fiber is then drawn back into the protective needle and the needle is withdrawn from the sample container. At this point the needle of the extraction assembly is injected into the sample port of a gas chromatograph. Then the fiber is extended and exposed to the heated injection chamber, causing the analytes to be desorbed from the fiber. The analytes are then focused at the inlet of the capillary GC column.

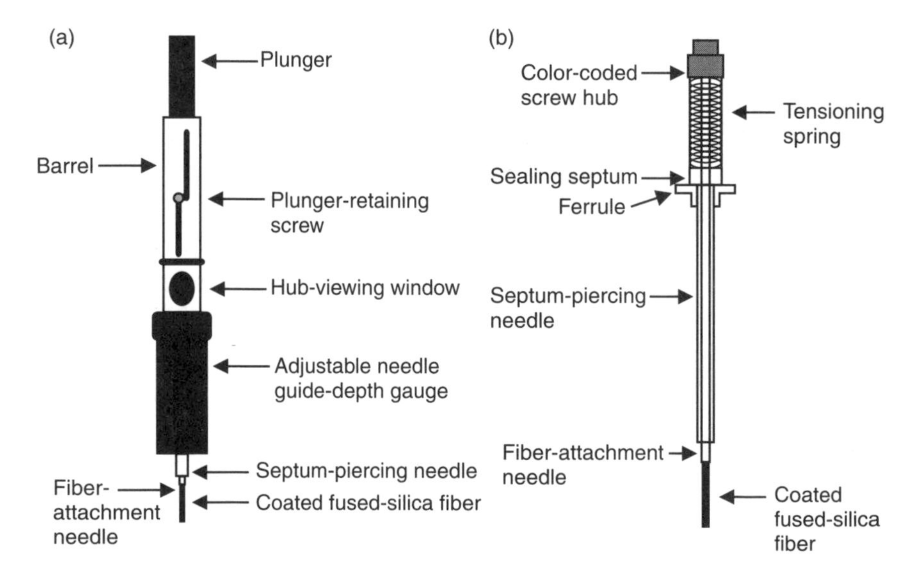


**FIGURE 8.1** Schematics of (*a*) the SPME device and (*b*) the fiber assembly. [From V. Mani and C. Wolley, *LC-GC* **13** (1995) 734 with permission.]

The SPME has several important advantages. The time required for the analyte to reach an equilibrium between the coated fiber and the sample is relatively short. The analytes are thermally desorbed rapidly and completely into a gas chromatograph, yielding sharp symmetric peaks. No solvent is required, so solvent disposal is eliminated. Researchers have obtained low parts-per-trillion (ppt) detection limits with electron-capture and ion-trap detectors (3). SPME can be used with any GC or GC-MS system with split, splitless, direct, or on-column sample injection.

Perhaps the most serious disadvantage of SPME is that it is an equilibrium technique. Often, only a small fraction of the sample analytes are extracted by the coated fiber. Quantification is dependent on extracting a precisely known fraction of each analyte. This means that a change in the sample matrix, or any other variable that affects the equilibrium, may affect the quantitative results. By contrast, conventional solid-phase extraction is usually a total-extraction technique, meaning that *all* of each analyte is transferred to the solid extraction phase.

### 8.1.2 Theory

The principle of SPME is the same as any extraction method, the partitioning of analytes between two phases. Most commonly, analytes of interest are extracted from a predominantly aqueous sample into an organic polymeric phase coated onto a silica fiber. The extent to which an analyte is extracted by the coated fiber will depend on the value of the mass distribution ratio, $D$:

$$D = K_d \frac{V_2}{V_1} \tag{8.1}$$

where $V_2$ is the volume of the coated phase, $V_1$ is the volume of the sample solution, and $K_d$ is the distribution coefficient. Here it is assumed that the analyte exists in only a single chemical form. (See Section 1.2.3 for a more complete discussion of extraction equilibrium constants.)

The fraction of an analyte extracted onto the coated fiber from the sample solution is given by

$$f_{ex} = \frac{D}{D+1} = \frac{K_d V_2}{K_d V_2 + V_1} \tag{8.2}$$

In SPME the ratio of the two phases ($V_2$, the coated phase; $V_1$, the sample liquid phase) is usually very small. Therefore, the fraction of an analyte extracted ($f_{ex}$) will depend primarily on the distribution coefficient ($K_d$) and on the volumes of the two phases, $V_2$ and $V_1$.

The fraction of an analyte extracted under various conditions can be illustrated by some examples. If a fiber 0.2 mm in diameter is coated to a length of 10 mm with a polymeric coating 50 μm thick, the coated area is $0.2\ \pi \times 10 = 6.3\ \text{mm}^2$ and the volume ($V_2$) is $6.3\ \text{mm}^2 \times 0.05\ \text{mm} = 0.31\ \text{mm}^3$. Substituting into Equation (8.2), we can calculate the fraction of solute extracted for a sample ($V_1$) of 1.0 ml ($1.0 \times 10^3\ \text{mm}^3$) and for $V_1 = 10$ ml ($10 \times 10^3\ \text{mm}^3$) assuming different values for the distribution coefficient, with percentage extraction = $100 f_{ex}$:

| $K_d$ | $f_{ex}$ for $V_1 = 1.0$ ml | $f_x$ for $V_1 = 10$ ml |
|---|---|---|
| 100 | 0.03 | 0.003 |
| 500 | 0.13 | 0.015 |
| 1,000 | 0.24 | 0.03 |
| 5,000 | 0.61 | 0.13 |
| 10,000 | 0.76 | 0.24 |

These examples show that often a relatively small percentage of the analyte is actually extracted. However, quantitative work is based on a constant fraction of an analyte being extracted. The mass ($n$) of substance extracted is equal to its initial concentration ($C_0$) times the fraction extracted, $f_{ex}$:

$$n = C_0 f_{ex} \tag{8.3}$$

Thus, if the final analysis is done by gas chromatography the GC signal will be a linear function of concentration ($C_0$), provided $f_{ex}$ is the same for all concentrations within the analytical range. This will usually be the case unless there is a major change in the sample matrix that will affect the distribution ratio of an analyte and thereby change the value of $f_{ex}$. In SPME the entire amount of the extracted analyte ($n$) is desorbed into a GC. This being the case, the GC detection limit is usually desirably low even though $f_{ex}$ may be considerably less than 1.0. The GC peak height or area is of course a function of the amount of an analyte ($n$) that is injected into the instrument. Equation (8.3) predicts that a plot of peak height or area against the concentration of analyte in the sample ($C_0$) will be linear with a slope equal to the fraction extracted, $f_{ex}$. A higher $f_{ex}$ gives a higher slope and should lead to better precision and accuracy. The dependence of chromatographic peak height on $f_{ex}$ as well as sample concentration also means that a calibration curve will be required for each analyte (unless $f_{ex}$ is already known accurately).

### 8.1.3 Kinetics

For aqueous samples, the rate at which partition equilibrium is achieved depends primarily on the rate at which the analytes reach the surface of the coated fiber. Diffusion will take longer from samples of larger volume. Vigorous agitation methods, such as sonication, can greatly speed up the equilibration process (4). The equilibrium time is much longer (2–15 min) for more practical agitation methods such as magnetic stirring. One of the best ways to overcome the kinetic limitation is to heat the sample.

Louch et al. developed a model to describe the rate at which adsorbed organic analytes reach equilibrium in a perfectly agitated liquid solution (5). In this model diffusion rate differences within the coating were assumed to limit the rate of analyte uptake by the fiber. However, on the basis of experimental results it was concluded that in agitated solutions the rate of uptake was controlled by diffusion through a thin stagnant aqueous layer around the fiber. As this was the case, it was argued that diffusion differences within the coated fiber could be neglected.

Using fibers with a polyacrylate coating, Vaes et al. showed that with efficient agitation it is possible to achieve a sufficiently small aqueous diffusion layer to prevent diffusion through this layer from being the limiting factor in the adsorption equilibrium process (6). This implies that the equilibrium time is determined solely by the polyacrylate phase. These authors also concluded that hydrophobicity and hydrogen bonding play significant roles in partitioning of analytes between aqueous samples and polyacrylate coatings.

At this point it must be emphasized that efficient sample agitation is essential to attain a reasonably fast equilibrium. Without any stirring, sample analytes reach the fiber surface only by diffusion through the aqueous solution, which is a relatively slow process.

It will be seen from Equation (8.2) that a thicker coating on the fiber ($V_2$) will result in a greater uptake of analytes. Slower diffusion within the thicker coating will increase somewhat the time for equilibrium. However, this may have an adverse effect on desorption; it simply takes longer to desorb all the analyte into the GC. This creates more opportunity for precolumn band broadening. It can also result in carryover from one run to the next when the fiber is used for multiple runs. Very high distribution ratios values can also create a carryover problem because so much analyte is extracted by the coating. Potter and Pawliszyn discovered this phenomenon using PDMS-coated fibers (7). Benz[*a*]anthracene, for example, had "significant" carryover. In an attempt to solve this problem, the fiber was desorbed a second time to drive off the residual analyte.

The overall result is something of a paradox; high values of the equilibrium constant are desired for maximum extraction, but carryover can become a problem. Desorption of extracted analytes by a small volume of a liquid solvent has been used, but this tends to defeat a major attribute of SPME, that of direct thermal desorption into a gas chromatograph.

### 8.1.4 Experimental Variables

Several experimental parameters have been studied with a goal of improving the speed, sensitivity, and selectivity of SPME. Attention to seemingly small details can often contribute greatly to the success of an analytical technique.

***8.1.4.1 Agitation Methods*** Magnetic stirring is probably the most convenient method for agitation of the small liquid samples used in SPME. Pawliszyn (8) found that star-shaped stir bars worked better than disk or cylindrical bars. Speeds of up to 3300 rpm (rev/min) were possible, resulting in an equilibrium time of only 2 min for 0.10 ppm of benzene or toluene.

Intrusive stirring methods with an Osterizer blender permitted speeds of up to 13,100 rpm. Equilibrium time was reduced to 100 s. Sonication was even faster, providing an equilibrium time of only 50 s. However, the equipment used required water cooling (8).

***8.1.4.2 Coating Materials*** The selectivity of a coated fiber for various analytes can be enhanced by choosing a coating similar in chemical structure to that of the analyte. The following examples illustrate some of the useful coating materials:

- *Poly(dimethyl)siloxane.* This nonpolar coating has been used for SPME of alkylbenzenes and PAHs (9), volatile halogenated compounds (10) and flavor compounds (11,12)
- *Mixed Polymers.* Among other compounds, Carbowax and poly-divinylbenzene have been used for SPME of alcohols and small polar compounds (14)

Increasing the thickness of the coating material will increase the volume ($V_2$) and extract a higher proportion of the analyte. However, the price of this improvement is a longer equilibration time. In one study (4), a 97-μm-thick coating of poly(dimethyl)siloxane extracted 80 ng of benzene from 1 ppm solution in 200s, a 56-μm coating extracted 40 ng in 100 s , and a 15-μm coating extracted 10 ng of benzene in 30 s.

***8.1.4.3 Salting-out Effect*** Addition of an inorganic salt to the aqueous sample shifts the partition equilibrium so that more of the analytes will be extracted. A substantial concentration of salt must be used for this technique to be effective. Salt concentration has little effect up to approximately 1% salt. This means that SPME can be used either in fresh or brackish water samples without recalibration.

Added salt resulted in a large enhancement in extraction of phenols at pH 2 where the phenols were present as neutral molecules (15). At pH 7 the phenols were partially ionized and added salt reduced the amounts extracted.

***8.1.4.4 Effect of pH*** As with any extraction method, sample pH can be adjusted to provide better selectivity in SPME. The coating material extracts only compounds that are in molecular form; ions generally are not extracted. An acidic sample pH results in a large enhancement in the extraction of phenols that have a relatively low $pK_a$ value.

***8.1.4.5 Sample Heating*** Heating liquid samples results in faster diffusion rates of the analytes to the coated capillary surface. This can result in a very significant reduction in the time needed for equilibrium. However, at higher temperatures less analytes is extracted because the extraction process is exothermic.

Heating of gaseous samples is not desirable because diffusion rates are already fast and equilibrium times are short.

***8.1.5 Headspace SPME*** Two basic methods are used for SPME. One is by direct immersion of the fiber into the sample. The sample may be a liquid, a soft solid containing a liquid, or a gas if air samples are being analyzed. The other basic method is to sample the headspace above the sample in an enclosed container. Use of the headspace mode can avoid one of the major problems of direct-immersion SPME. If the sample is dirty, such as from a muddy stream or a sewage sample, the fiber coating can be plugged by the sample solids. In such cases, SPME will work no better than other SPE methods.

Sampling the headspace over the sample can avoid the problem of plugging the coating by solids. Arthur and Pawliszyn (16) demonstrated this technique with the BTEX suite of compounds: benzene, toluene, ethylbenzene, and xylenes. Standard samples of these components demonstrated that the equilibrium constant of the fiber with the analytes in the solution was a product of the water-to-air and air-to-coating partition coefficients. An equation for the fraction of analyte transferred from the aqueous sample to the coating in headspace can be derived as follows:

Water-to-air equilibrium:

$$K_1 = D_1 \frac{V_1}{V_2} \tag{8.4}$$

where $V_1$ = volume of water and $V_2$ = volume of air.

Air-to-coating equilibrium:

$$K_2 = D_2 \frac{V_2}{V_3} \tag{8.5}$$

where $V_3$ = volume of coating.

Water-to-coating equilibrium:

$$K_1 K_2 = D_1 D_2 \frac{V_1}{V_3} = D_3 \frac{V_1}{V_3} \tag{8.6}$$

The total fraction of analyte extracted by the coating ($f_{ex}$) is given by the expression

$$f_{ex} = \frac{D_3}{D_3 + 1} = \frac{K_1 K_2 V_3}{K_1 K_2 V_3 + V_1} \tag{8.7}$$

Equation (8.7) is identical in form to Equation (8.2), but it has an additional $K$ term.

Headspace may also be used to sample volatile components released from solid samples. The overall extraction in headspace with either liquid or solid samples is apt to be lower than in direct-immersion SPME because transfer of analytes from the sample to the gas phase seldom is quantitative. Heating the sample will usually increase the total extraction by causing a greater fraction of the analytes to be volatilized.

An obvious drawback to headspace SPME is that only the more volatile compounds can be determined; semi- and nonvolatiles, which could equilibrate with the fiber in direct immersion, will not be present in detectable amounts in the headspace. Headspace SPME has obvious utility for the collection and analysis of a wide variety of volatile substances. For example, this technique can be used for qualitative identification of natural volatile substances from plants, fungal cultures, insects, and other sources (9). However, quantification of these volatiles is not trivial. Each volatile substance will have a different equilibrium partition constant between the headspace and the SPME fiber, so that the relative GC peak areas do not reflect the true proportion of these analytes in the headspace. Other factors such as sampling time and temperature will also affect the quantification. One answer to this problem is to calculate calibration factors (amount adsorbed by the fiber divided by the headspace concentration) from the Kovats retention indices of the volatile compounds (17). A regression model has been developed for predicting the calibration factors from the GC retention index, temperature, and analyte functional class.

In a related study not involving headspace, a method for estimating distribution coefficients for hydrocarbons between air and SPME-coated fibers was developed (18). The distribution coefficients were calculated directly from gas chromatographic data using linear temperature–programmed retention index system (LT-PRI). There is a linear relationship between log $K$ and LT-PRI for a series of normal alkanes. This approach makes it possible to quantify hydrocarbons in air exclusively on the basis of chromatographic parameters without any calibration.

### 8.1.6 Applications

The following applications have been selected to give a feeling for the scope and practical utility of SPME.

***8.1.6.1 Phenols*** In their early work, Pawliszyn and co-workers demonstrated that SPME could be used to extract phenols from aqueous samples (19). Eleven different phenols were studied using a fiber coated with a 95-μm layer of polyacrylate. The extractive fiber was placed in a 40-ml aqueous sample with pH adjusted to 2.0. Equilibrium was attained within 40 min. High salt concentrations led to longer equilibrium times.

***8.1.6.2 PAH Compounds*** SPME of polyaromatic hydrocarbons (PAH) and polychlorinated biphenyls was performed using a 15-μm polydimethylsiloxane-coated fiber (20). The following compounds were included in the study: naphthalene, anthracene, benz[*a*]pyrene, benz[*a*]anthracene, PCB3, and PCB5. Each analyte was spiked into deionized water, and the fiber was exposed to the solutions for various time periods. Each solution was sonicated during the extraction/preconcentration step to provide agitation and reduce the time needed for equilibrium. It was found that after an exposure time of only 10 min, each analyte was recovered linearly from 6.25 pg/ml to 3.75 ng/ml with the worst sensitivity reported for naphthalene due to its low *K*(octanol/water) value. Further testing resulted in estimated limits of detection well below those mandated by EPA methods. For the 10-min extraction time, the following estimated limits of detection were reported: 1.25 pg/ml for benz[*a*]pyrene and benz[*a*]anthracene, 125 pg/ml for naphthalene (estimated), 2 and 3 pg/ml for the rest of the compounds.

Subsequently, the authors demonstrated the application of the method to real-world samples. In the first example, a groundwater sample, obtained from a Canadian military base, was examined. The SPME method was coupled with GC/ITMS, and the sample was found to contain naphthalene, methylnaphthalene, biphenyl, acenaphthalene, dibenzofuran, and 9*H*-fluorene. Using the SPME method, the naphthalene concentration was found to be 15 μg/mL in the sample. This compared favorably to a conventional liquid–liquid extraction value of 19 μg/ml. Finally, the recoveries of the selected, deuterated analyte compounds in sewage water were determined. Good recoveries (>90%) were reported for those compounds with relatively low *K*(octanol/water) values, such as 1,4-dichlorobenzene, naphthalene, and acenaphthalene. Compounds with relatively high *K*(octanol/water) values were found to give very poor recoveries (8–34%), and it was postulated that significant matrix problems were present in the sewage samples. Most likely, suspended organic matter is competing with the sorbent on the fiber for the analyte.

While successfully demonstrating the usefulness of SPME on a relatively clean sample, as in the case of the groundwater sample, the main limitation of the technique becomes apparent in these examples because of the incomplete extraction of the target analytes. It would appear, then, that in the case of calibration and absolute recovery, SPME is restricted in its ability to deal with variations in the sample matrix.

***8.1.6.3 Volatile Organic Compounds (VOCs)*** SPME has been used to determine VOCs and gasoline components in groundwater and surface water sample (22,23). Parts of this work involved the use of polydimethylsilane (PDMS) chips with a volume or approximately 80 μl. After the extraction step, the chips were placed in a UV cell with deionized water and centered, and the adsorbed analytes was measured by UV absorption.

Because this measurement is nondestructive, the chips could then be placed in a special adapter for normal desorption into a gas chromatograph. This provides a double check. In contaminated real water samples the sensitivity was as good as in pure water. Thus, suspended particles and other materials did not interfere.

Nilsson et al. (23) determined 60 volatile compounds in water samples. Water-soluble components of jet fuels were also determined (24). Owing to the high use of jet fuels at military installations, measurement of the compounds detected in water can aid in determining both the quantity and type of fuels spilled into the ground over the years. SPME is an attractive technique for such measurements, especially for screening in the field.

***8.1.6.4 Pesticides*** Another large area of concern is the determination of pesticides and their residues in groundwater and surface water. Page and Lacroix (25) employed a headspace method for semivolatile compounds with heating to 87 °C, Heptachlor, aldrin, and *cis*-permethrin, for example, all showed an increase in sensitivity of a factor of at least 10 over current EPA methods. Young et al. (26) showed that using the headspace or direct-immersion methods met or exceeded the EPA method now used for 20 organochlorine compounds. They did, however, encounter problems with carryover. This appeared to be because of the direct-immersion technique and the length of time for equilibration. The GC simply could not desorb all of the analyte.

Finally, Boyd-Boland et al. (27) determined mixtures of as many as 60 pesticides. Mixtures of all 60 did not appear to affect how well each compound extracted. All were determined at similar levels whether determined simultaneously or individually. These compounds contained nitrogen, chlorine, and phosphorus groups within their structures and all were detected at limits of nanograms per liter to subnanograms per liter. In a comparison with liquid–liquid extraction (LLE) for heptachlor from a real water sample, SPME gave 1.84 mg/kg while LLE gave 1.85 mg/kg.

***8.1.6.5 Analysis of Human Breath*** More than 100 compounds have been identified in normal human breath by gas chromatography and mass spectrometry (28). Breath analysis can be used as a diagnostic tool because increases or decreases in the concentrations of certain compounds have been associated with various diseases or changes in metabolism (29). Grote and Pawliszyn have successfully applied SPME for the quantitative determination of ethanol, acetone, and isoprene in human breath (30). The equipment used was a SPME device modified for breath analysis. Three different fiber coatings were evaluated with regard to sensitivity, linear range, precision, and detection limits. Changes in temperature and humidity will, of course, affect the partition coefficients, so these variables were studied in detail.

***8.1.6.6 Foods and Beverages*** Various chemicals in food samples can be determined by inserting the SPME probe into a solution of the dissolved food, an aqueous slurry of the food, or even into a soft solid food. Conventional SPE filtration devices would most likely be plugged by the food solids. However, it should be kept in mind that the chemicals may partition between the suspended solids and water as well as between water and the coated probe.

Their complication may be partly avoided by the use of headspace in some instances. Halogenated volatile organic compounds in selected foods have been measured by headspace (10). Caffeine has been determined in beverages in direct-immersion SPME (21).

### 8.1.7 Conclusion

Any method of analysis has its limitations. As with any method, the limitations of SPME can be minimized with patience, practice, and ingenuity. Because this technique is so new, the paths around these limitations have been barely explored. The advantages that SPME can provide are enticing: little, if any, liquid solvents are used; the method is fast and inexpensive; and the apparatus is readily portable for field sampling.

## 8.2 SEMIMICRO SOLID-PHASE EXTRACTION (SM-SPE)

### 8.2.1 Introduction

Membranes require less eluting solvent and are generally more efficient than cartridges and mini columns containing loose sorbent particles. However, the full advantages of membranes for SPE on a semimicroscale have yet to be realized. Large membrane disks require up to 10 ml of eluting solvent and packed minicolumns often require 1–2 ml. Of this, only a small fraction (1–2 μl) is generally used for measurement of extracted analytes by gas chromatography. To obtain better sensitivity and detection limits, eluates are often evaporated down to a fraction of the original volume of eluate. This additional step reduces the speed of SPE and risks the chance of some loss of sample.

A technique has been described for performing solid-phase extractions on a semimicroscale (31). Thin membrane disks 4 mm in diameter containing lightly sulfonated polystyrene or Silicalite particles are placed in the hub of a syringe needle. Aqueous samples (1–6 ml) are passed through the membrane disks, and the extracted compounds are subsequently eluted with 20–50 μl of an organic solvent. Unlike solid-phase microscale extraction (SPME), which uses a coated fiber, this method is essentially a total-extraction technique. Recoveries >90% were generally obtained for a wide variety of test compounds.

### 8.2.2 Extraction Device

The apparatus used for SM-SPE is shown in Figure 8.2, which is drawn larger than actual scale to show the details better. The apparatus consisted of a Hamilton 1000 series Gastight 5-ml glass syringe with a Teflon luer lock (Catalog No. 81520) acting as the sample reservoir. The syringe was fitted with fluorocarbon-hubbed, 22-gauge stainless steel needles (Catalog No. 90134), which served as the extraction columns. The bottom of each needle was gently tapered to a point by the machine shop. A piece of

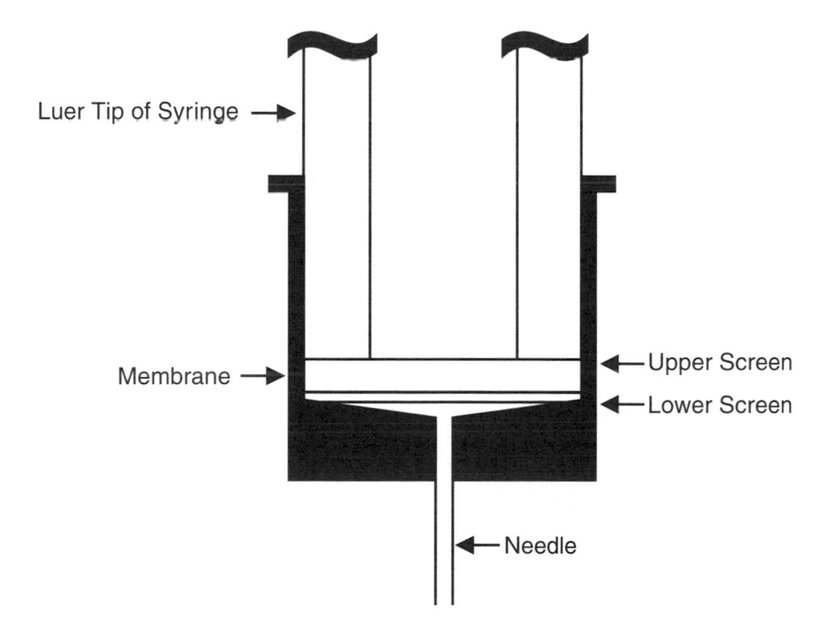


**FIGURE 8.2** Device used for SM-SPE. This figure has been drawn approximately to scale and enlarged for clarity. (From Ref. 23 with permission.)

stainless steel wire mesh, 4-mm in diameter, 228 μm thick, and 53.3% open pore volume, was machined into place just above the tapered bottom to support the membranes.

The device used for SM-SPE (Fig. 8.2) must be carefully designed to provide efficient extraction and subsequent desorption on a small scale. The $0.5 \times 4$-mm membrane disks weigh only ~9 mg (PS-DVB) and 11 mg (Silicalite), of which 90% is estimated to be solid particles for the extraction. The disks are rather soft and need to be supported on a thin stainless steel screen that is machined into place in the needle hub. It was necessary to drill out the inside of the needle hub slightly so that a conical void was created just below the membrane disk and its mesh support. This was needed to assure a smooth flow of liquid through the membrane and out the needle. Without the tapered void, the disk lay flat against the needle hub, causing the backpressure to rise and the volume of solvent needed for elution to double.

In the work described, the needle containing a small membrane disk was connected to a small syringe. The aqueous sample was placed in the syringe barrel and forced manually through the needle assembly. Alternatively, air or gas pressure could be used to push through the sample.

The procedure used for SM-SPE was as follows (31). The plunger was removed from the syringe barrel and a packed needle locked into place. Samples were prepared by adding a 10-μl aliquot of methanol solution containing 100 ppm each of two to three analytes to 1–6 ml of deionized (DI) water. The final concentration of each analyte in the sample was 0.17–1.0 ppm.

After loading was complete, the syringe barrel was rinsed with approximately 200 μl DI water and air was pushed through the membranes to remove any remaining water. A 20-50 μl aliquot of acetone, ethyl acetate, or methylene chloride was used to elute the compounds into a capped GC vial. An internal standard, 2 μl of acetone solution containing 250 ppm toluene, was added to the contents of the vial, which were analyzed by gas chromatography using manual injection.

### 8.2.3 Scope and Results

Experimental Empore-type membranes containing lightly sulfonated PS-DVB resins have been shown previously to effectively extract a wide variety of organic test compounds from aqueous samples (31). Excellent recoveries were obtained with these membranes on a semimicroscale, as indicated by the data in Table 8.1. The aqueous samples contained 0.17–1.0 ppm each

**TABLE 8.1 SM-SPE Recovery of Organic Compounds Using a Sulfonated PS-DVB Resin-Loaded Membrane**

| Class | Compound | Volume Desorption Solvent, μl | Average Recovery, % | |
|---|---|---|---|---|
| | | | 1-ml Sample | 6-ml Sample |
| Phenol | Phenol | 40 | 92 | 91 |
| | *o*-Cresol | 40 | 102 | 102 |
| | 2,5-Dimethylphenol | 40 | 101 | 98 |
| | 2-Chlorophenol | 40 | 100 | 93 |
| | 4-Chlorophenol | 40 | 92 | 98 |
| | 3-Nitrophenol | 40 | 94 | 93 |
| Aldehyde | *n*-Valeraldehyde | 40 | 88 | 90 |
| | Octaldehyde | 50 | 89 | 86 |
| | Benzaldehyde | 50 | 99 | 98 |
| | Salicylaldehyde[a] | 40 | 100 | 95 |
| Alcohol | 1-Pentanol[b] | 20 | 102 | 95 |
| | 3-Phenyl-1-propanol[a] | 20 | 92 | 91 |
| | 2-Ethyl-1-hexanol | 20 | 94 | 94 |
| | 1-Octanol[b] | 20 | 97 | 96 |
| Ester | Ethylacetoacetate | 20 | 99 | 97 |
| | Hexylacetate | 40 | 99 | 98 |
| | Methylbenzoate | 40 | 92 | 91 |
| | Isopentylbenzoate | 30 | 98 | 93 |
| Ether | Anisole | 50 | 98 | 98 |
| Ketone | 2-Pentanone | 30 | 98 | 93 |
| | 4-Methyl-2-pentanone | 30 | 100 | 94 |
| | 2-Hexanone[b] | 30 | 100 | 100 |
| | Acetophenone | 40 | 99 | 99 |

[a]Eluted with methylene chloride.

[b]Eluted with ethyl acetate.

*Source:* From Reference 31 with permission.

of several test compounds. The average recovery for all test compounds was 97% for 1-ml samples and 95% for 6-ml samples. The relative standard deviation was 1.7%.

The volume of acetone required for elution varied from 20 to 50 μl. In some cases methylene chloride or ethyl acetate was used for the desorption step. With a 20-μl injection into a gas chromatograph, the fraction of the organic solvent eluate actually injected into the GC varies from about 0.1

to 0.04. In conventional SPE the fraction of eluate injected is often on the order of 0.001.

In Chapter 6 it was shown that sulfonated PS-DVB membranes extract most organic compounds well but that experimental Silicalite-loaded membranes are able to better extract small, polar organic compounds. In SM-SPE it should be feasible to use two membranes of different types in the needle hub. This proved to be very successful, as demonstrated by the data in Table 8.2. The membrane packing order did not influence the recoveries. The first analyte in each pair was selectively extracted by the Silicalite-loaded membrane and extracted little, if at all, by the sulfonated PS-DVB resin-loaded membrane. The second analyte in each pair was more favorably extracted by the sulfonated PS-DVB resin-loaded membrane. All recoveries were about 90% or greater with a relative standard deviation of 2.3%. These results show that the membranes perform just as well in tandem as they do alone. This technique promises to expand the scope of organic compounds that are amenable to solid-phase extraction.

### 8.2.4 Double-Pass Sampling

A convenient manual technique for SM-SPE is to draw the sample via the needle tip up through the membrane. The sample is then pushed back through the membrane a second time and expelled out the needle. This double-pass technique should ensure a high degree of extraction. It is

**TABLE 8.2 SM-SPE Using Mixed Membranes[a]**

| Compound Pair | Recovery, % |
|---|---|
| Methylacetate, methylbenzoate | 91, 95 |
| 2-Butanol, 2-cresol | 99, 102 |
| 1-Butanol, 3-nitrophenol | 94, 99 |
| Ethyl butyrate, 2,5-dimethylphenol | 90, 93 |
| Ethyl acetate,, silicylaldehyde | 90, 90 |
| 2-Pentanone, 2-chlorophenol | 94, 100 |
| Ethylpropionate, Phenol | 96, 93 |

[a]*Conditions:* 0.5-mm-thick sulfonated PS-DVB resin-loaded and 0.5 mm-thick Silicalite-loaded membranes, single-pass, 6-ml aqueous samples, 1 ml/min. Elution with 40 μl acetone, 10 s. Analyte concentration 0.083 ppm.

*Source:* From Reference 31 with permission.

particularly convenient for field sampling; elution of the analytes immobilized on the membrane can be completed in the laboratory if desired.

Minor adjustments in the membrane–needle assembly were required for the double-pass technique to be practical. A second mesh screen was inserted above the membrane to hold it in place during the sample-draw step. Sampling was slow during the draw step due to the resistance of flow of 1-mm-thick membranes. This problem was avoided by reducing the membrane thickness to 0.5 mm or in some cases to 0.33 mm.

Recovery data are given in Table 8.3 for samples extracted by a $0.5 \times 4$-mm sulfonated PS-DVB resin-loaded membrane using double-pass sampling. The sampling time for a 1-ml aqueous sample was only 45–60 s (draw time 30–45 s; push time 15 s). The average recovery for sample compounds was 89% with a RSD of 2%. Recovery data for double-pass extractions using a $0.33 \times 4$ mm Silicalite-loaded membrane are given in Table 8.4. In this case the sampling time was 50–85 s (draw time 30–45 s; push time

**TABLE 8.3 Recovery of Test Compounds Using a Sulfonated PS-DVB Resin-Loaded Membrane (0.5 mm Thick) and Double-Pass Sampling[a]**

| Class | Compound | Recovery, % |
|---|---|---|
| Phenol | *o*-Cresol | 60 |
| | 2,5-Dimethylphenol | 76 |
| | 4-Chlorophenol | 68 |
| Aldehyde | Octylaldehyde | 100 |
| | Benzaldehyde | 95 |
| | Salicylaldehyde | 100 |
| Alcohol | 3-Phenyl-1-propanol | 85 |
| | 2-Ethyl-1-hexanol | 93 |
| | 1-Octanol | 99 |
| Ester | Hexylacetate | 100 |
| | Methylbenzoate | 94 |
| | Isopentylbenzoate | 100 |
| Ether | Anisole | 81 |
| Ketone | 2-Hexanone | 82 |
| | 2-Heptanone | 100 |

[a]*Conditions:* 1.0-ml aqueous samples containing 0.5 ppm of each test compound. Draw time 30–45 s, push time 15 s. Elution with 20–50 μl acetone in 5–10 s.

*Source:* From Reference 31 with permission.

**TABLE 8.4 Recovery of Test Compounds Using a Silicalite-Loaded Membrane (0.33 mm Thick) and Double-Pass Sampling[a]**

| Class | Compound | Recovery, % |
|---|---|---|
| Alcohol | 1-Butanol | 86 |
| | 1-Pentanol | 100 |
| | 2-Pentanol | 93 |
| | 1-Hexanol | 103 |
| | 2-Ethyl-1-hexanol | 95 |
| | 1-Octanol | 101 |
| | 2-Octanol | 103 |
| Aldehyde | *n*-Valeraldehyde | 102 |
| | Hexaldehyde | 99 |
| | Benzaldehyde | 100 |
| Ketone | 2-Butanone | 92 |
| | 2-Pentanone | 103 |
| | 4-Methyl-2-pentanone | 102 |
| | 2-Hexanone | 103 |
| | 3-Hexanone | 99 |
| | 2-Heptanone | 100 |
| Ester | Ethylpropionate | 102 |
| | Ethylbutyrate | 101 |

[a]*Conditions:* 1-ml aqueous samples, 0.5 ppm of each test compound. Draw time 30–45 s, push time 20–30 s. Elution with 40 μl acetone, 10 s.

*Source:* From Reference 31 with permission.

20–30 s). The average recovery for sample compounds was 99% with a RSD of 2%.

The amount of sorbent particles in these thinner membranes is quite small: 4.5 mg for 0.5 × 4-mm sulfonated PS-DVB and ~3.7 mg for 0.33 × 4-mm Silicalite. However, the analyte in a 1.0-ml sample containing 1 ppm is smaller yet: ~1 μg. With such small amounts of solid–extractant particles in the membranes it is useful to know the effect of analyte concentration in the sample on its percentage recovery in SM-SPE. Recoveries of 2-hexanone by double-pass SM-SPE with a 0.33 × 4-mm Silicalite-loaded membrane were as follows: 0.005 ppm, 91%; 0.01 ppm, 96%; 0.10 ppm, 103%; 1.0 ppm, 106%; 10 ppm, 84%; and 100 ppm, 81%. Thus, recoveries >90% were obtained for 2-hexanone concentrations ranging from 0.005 to

1.0 ppm. Some overloading and, therefore, lower recoveries occurred at 10 and 100 ppm. Recoveries of 2-hexanone on the 0.5 × 4-mm PS-DVB resin-loaded membrane were lower: 0.005 ppm, 67%; 0.01 ppm, 70%; 0.10 ppm, 72%; 1.0 ppm, 79%; 10 ppm, 67%; and 100 ppm, 54%.

## 8.3 MINIATURIZED SOLID-PHASE EXTRACTION

### 8.3.1 Introduction

Essentially total extraction of analytes with resin-loaded membranes can be performed on an even smaller scale than that described in Section 8.2. A device dubbed *miniaturized solid-phase extraction* (M-SPE) employs an Empore membrane only 0.7 mm in diameter and 1.2 mm thick housed inside the needle chamber of a 50-μl syringe (32). Employing a membrane of this size reduces the sample size to the 0.5–2.5-ml range and requires only 5–10 μl of organic solvent for complete elution of the adsorbed analytes. A 2.0-μl aliquot of the eluate was injected into a gas chromatograph for subsequent analyses, or the entire eluate could be injected directly into a gas or liquid chromatograph. Recoveries of analytes from aqueous samples at the 10-ppb concentration level averaged >90% with good reproducibility. Concentration factors of at least 500:1 were obtained (32).

### 8.3.2 Extraction Assembly

A 50-μl Hamilton gastight syringe with removable needle (Model 1705) (Hamilton Co., Reno, NV) was modified to perform miniaturized SPE. The actual SPE apparatus housed inside the removable needle chamber is illustrated in Figures 8.3 and 8.4. To prepare the syringe for SPE, the top half of the needle inlet of a 22S-gauge needle was removed and a 0.25-mm-deep cone was machined. Positioned above this cone was a 55-μm stainless steel mesh screen on which the membrane rested. The dimensions of the membrane were 0.7 mm in diameter × 1.2 mm in height. All of these components were held in place by a Teflon ferrule that is a standard part of the Hamilton removable needle syringe series.

Membrane plugs were cut from larger Empore disks by means of a cutting device consisting of a small metal tube (0.7 mm ID) sharpened on the cutting end. Two plugs were cut from 0.6-mm-thick Empore disks, and the cutting device was inserted into the ferrule. The two plugs were then pushed into place by a wire plunger inside the cutting tube.

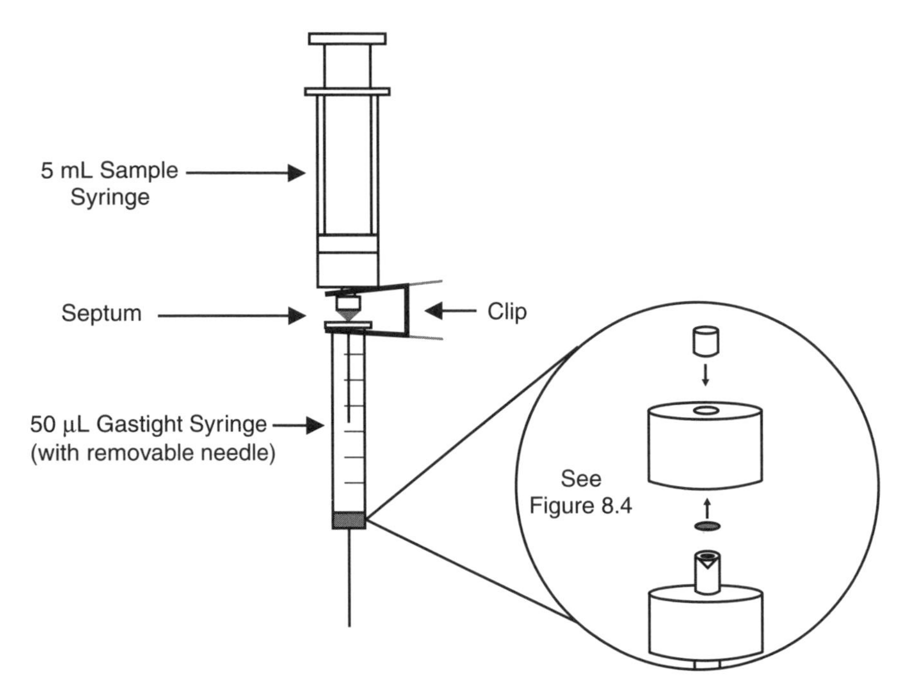


**FIGURE 8.3** Sample syringe and M-SPE syringe system.

Samples to be analyzed were added to a 5-ml syringe that was attached to the M-SPE syringe by a modified clip (see Fig. 8.3). A gray GC septum was placed between the two syringes to provide a pressurized seal and prevent any leakage when pressure was applied.

### 8.3.3 M-SPE Procedure

The extraction plug was initially cleaned by passing 20 μl of acetonitrile and then 35 μl of water through the system. An aqueous sample of appropriate size (0.5–2.5 ml) was loaded into the upper syringe and passed through the system by manual pressure applied to the syringe plunger. However, sample extractions could be partially automated by placing a Series 74900 single-syringe infusion pump (Cole-Palmer Instrument Co., Vernon Hills, IL) next to the upper syringe plunger. The motorized worm drive of the infusion pump pushed the sample through the M-SPE device at a flow rate of 100 μl/min. After passing the sample, the syringes were uncoupled and any remaining water was manually expelled from the M-SPE syringe.

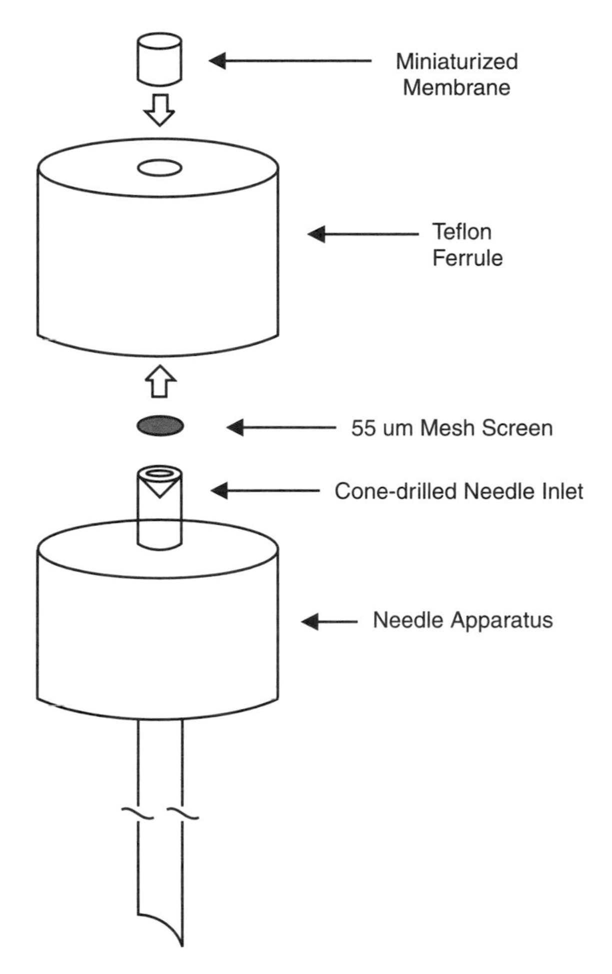


**FIGURE 8.4** Miniaturized-SPE assembly.

Extracted analytes were eluted from the miniature membrane by drawing up 5 μl of acetonitrile containing 5 ppm of mesitylene as an internal standard into the M-SPE syringe. This eluting solution was kept in contact with the membrane for 2–5 min, then pushed by manual pressure into a lined autosampler vial for GC analysis. In some cases direct injection into a gas chromatograph was used.

### 8.3.4 M-SPE of Substituted Benzenes

Empore membranes containing 8 μm of lightly sulfonated PS-DVB resins have already been shown to achieve high recoveries of a variety of organic solutes in SPE (33). However, in that study 1.0 ml of ethyl acetate was used to desorb the extracted analytes. Injection of a typical aliquot of 2 μl means that only 2/1000 of the analytes was used for the GC analysis. By contrast, injection of a 2 μl aliquot in M-SPE represents two-fifths or 40% of the extracted analytes.

To demonstrate that quantitative extraction data can be obtained on a very small scale with greatly reduced solvent consumption, the M-SPE syringe was used to preconcentrate substituted benzenes from aqueous samples (32). The membrane used again contained 8-μm lightly sulfonated PS-DVB resins.

The studies were conducted on 2.5-ml aqueous samples containing 10-ppb concentrations each of eight substituted benzenes. Once the sample loading onto the membrane was complete, 5 μl of internal standard (mestitylene) in acetonitrile was used to elute the concentrated analytes into an autosampler vial, from which 2-μl manual GC injections were performed.

Table 8.5 shows the average recoveries of four trials for a 500-fold preconcentration of seven substituted benzenes. The average recovery for these experiments was 92% with a relative standard deviation of ±4.6%. Excellent chromatographic peaks were obtained for all the sample compounds except for toluene. The toluene peak was bracketed by small peaks of impurities concentrated from the sample water.

### 8.3.5 Extraction Capacity of M-SPE

The volume of the membrane (0.7 mm in diameter, 1.2 mm thick) is calculated to be only 0.46 $mm^3$. A reasonable dead volume would be ~0.3-$mm^3$ (0.3 μl). This means that in the elution step at least an additional 0.3

**TABLE 8.5 Percent Recoveries of Substituted Benzenes after 500-Fold Concentration**

| Compound | ($n = 4$) Recovery, % |
|---|---|
| Benzene | 83 |
| Chlorobenzene | 96 |
| Ethylbenzene | 96 |
| Anisole | 99 |
| Bromobenzene | 94 |
| Propylbenzene | 89 |
| Butylbenzene | 87 |

μl of eluting solvent would be needed to displace the liquid holdup. This is not excessive, but it does place a lower limit on the amount of solvent that must be used for complete elution of the sample components.

The membrane plug may be estimated from its volume to contain 400–500 μg of actual resin. A 2.5-ml sample containing eight compounds each at 10 ppb (10 ng/ml) gives a total sample mass of around 200 ng, or 0.2 μg. Thus the sample loading by the resin is (0.2/500) 100 or only 0.04%. Resins typically have a loading capacity of around 1.0% for strongly adsorbed analytes. This predicts that the membrane plugs have more than adequate loading capacity.

## 8.4 CONCLUSIONS

SPME and membrane SPE performed on a semimicro- or miniaturized scale are really complementary techniques. SPME is performed on a truly microscale and conveniently transfers all the extracted to a gas chromatograph for final separation and measurement. But SPME is an equilibrium technique in which the fraction of each analyte extracted depends on its distribution coefficient. Often only a small fraction of each analyte is actually extracted. Because of this, careful calibration is required for quantitative work.

The membrane techniques, semimicroscale SPE and M-SPE, are essentially total extraction methods. Although these techniques require elution of extracted analytes from the membrane by a liquid solvent (rather than thermal desorption in SPME), this step is accomplished quickly with only a very small amount of eluting solvent. Final separation and measurement of the analytes can be accomplished by GC or HPLC.

In short, we have two quick and convenient techniques that should delight an imaginative scientist.

## REFERENCES

1. R. G. Belardi and J. Pawliszyn, *Water Pollut. Res. J. Can.* **25** (1989) 179.
2. Z. Zhang, M. J. Yang, and J. Pawliszyn, *Anal. Chem.* **66** (1994) 844A.
3. B. D. Page and G. Lacroix, *J. Chromatogr.* **648** (1993) 199.
4. Z. Zhang and J. Pawliszyn, *Anal. Chim. Acta.* **284** (1993) 265.
5. L. Louch, S. Morlagh, and J. Pawliszyn, *Anal. Chem.* **64** (1992) 1187.
6. W. H. J. Vaes, C. Hamwijk, E. U. Ramos, H. J. M. Verhaar, and J. L. M. Hermens, *Anal. Chem.* **68** (1996) 4458.

7. D. W. Potter and J. Pawliszyn, *J. Env. Sci. Techol.* **28** (1994) 298.
8. S. Motlagh and J. Pawliszyn, *Anal. Chim. Acta* **284** (1993) 265.
9. Z. Zhang and J. Pawliszyn, *J. High Res. Chromatogr.* **16** (1993) 689.
10. D. Page and G. Lacroix, *J. Chromatogr.* **648** (1993) 199.
11. X. Yang and T. Peppard, *LC-GC* **11**(13) (1995) 882.
12. V. Mani and C. Wooley, *LC-GC* **9**(13) (1995) 734.
13. C. Arthur and J. Pawliszyn, *Anal. Chem.* **62** (1990) 2145.
14. R. Shirey, V. Mani, and M. Butler, *The Reporter* **5**(14) (1995) 4.
15. K. Buchholz and J. Pawliszyn, *Anal. Chem.* **66** (1994) 160.
16. C. L. Arthur and J. Pawliszyn, *Anal. Chem.* **62** (1990) 2145.
17. R. J. Bartlet, *Anal. Chem.* **69** (1997) 364.
18. P. A. Martos, A. Saraullo, and J. Pawliszyn, *Anal. Chem.* **69** (1997) 402.
19. K. D. Buchholz and J. Pawliszyn, *Env. Sci. Technol.* **27** (1993) 2844.
20. D. Potter and J. Pawliszyn, *Env. Sci. Technol.* **28** (1994) 298.
21. J. J. Langenfeld, S. B. Hawthorne, and D. J. Miller, *Anal. Chem.* **68** (1996) 144.
22. B. L. Wittkamp, S. B. Hawthorne, and D. C. Tilotta, *Anal. Chem.* **69** (1997) 1204.
23. T. Nilsson, F. Pelusio, B. Larsen, S. Facchetti, and J. O. Madsen, *HRC-J. High Resol. Chromatogr.* **18** (1995) 617.
24. J. Ritter, V. K. Stromquist, H. T. Mayfield, M. V. Henley, and B. K. Lavine, *Microchem. J.* **54** (1996) 59.
25. B. D. Page and G. Lacroix, *J. Chromatogr. A* **757** (1997) 173.
26. R. Young, V. Lopez, and W. F. Beckert, *HRC-J. High Resol. Chromatogr.* **19** (1996) 247.
27. A. A. Boyd-Boland, S. Magdic, and J. Pawliszyn, *J. Analyst* **121** (1996) 929.
28. B. K. Krotoszynski, G. Gabriel, and H. O'Neill, *J. Chromatogr. Sci.* **15** (1977) 239.
29. A. Manolis, *Clin. Chem.* **29** (1983) 5.
30. C. Grote and J. Pawliszyn, *Anal. Chem.* **69** (1997) 587.
31. D. L. Mayer and J. S. Fritz, *J. Chromatogr. A* **773** (1997) 189.
32. J. J. Masso, R. C. Freeze, and J. S. Fritz, unplublished results, 1998.
33. P. J. Dumont and J. S. Fritz, *J. Chromatogr. A* **691** (1955) 123.

# CHAPTER 9

# APPLICATIONS

## 9.1 INTRODUCTION

The true test of any analytical technique is how well it works for practical analyses where the samples may be much more complex and difficult to handle than the clean, idealized samples used in the initial evaluation of the method. In order for an analytical method to gain wide acceptance, there must be a perception that (1) there is a real need for the method and (2) the technique works well on actual samples.

The importance of sample preparation in the overall analytical process has finally been recognized. There is increasing pressure to do a large number of analyses very quickly and efficiently in areas such as drug screening and genome projects. SPE is frequently an essential ingredient in rapid analyses. The excellent performance of SPE on "real" samples is attested to by the large number of practical applications. A sampling of these will be given later in this chapter.

In order for a technique to be used extensively in most contemporary laboratories, it is essential to have convenient products that are available commercially. Fortunately, this is now the case. Cartridges suitable for SPE are very popular. Commercial suppliers now offer a wide variety of particles and equipment for SPE. Automated workstations have been developed to greatly speed up SPE when a large number of samples must be handled (see Chapter 4). New and more efficient forms of SPE have been devised. For example, resin-loaded membranes have a number of advantages over cartridges and columns packed with loose extraction particles (see Chapter 6). Solid-phase microscale extraction (Chapter 8) is conceptually a very simple form of SPE and can be used directly on semiliquid samples or even on some types of solid samples.

The intent of this chapter is to give the reader some feeling of the broad scope of SPE as applied to actual samples. References are given to a number of original papers to illustrate the analytical problems to which SPE has been applied and the analytical methods that have been used. No attempt has been made to present a complete or exhaustive literature coverage. One reason for this is that practical applications of SPE have become too numerous to catalog adequately. Another is that there is a great deal of similarity and repetition in the extractive methods.

Illustrative applications of SPE according to sample type are discussed in Section 9.2 The major SPE techniques are summarized in Section 9.3, and some special problems associated with practical SPE are considered in Section 9.4. In the final section illustrative references to practical applications of SPE are listed according to the chemical classification of the various analytes.

A growing number of scientists are applying solid-phase micro extraction (SPME) to a wide variety of analytical samples. The essentials of SPME, together with illustrative applications, are covered in Chapter 8.

## 9.2 SAMPLE TYPES

### 9.2.1 Sample States

Liquid samples, or samples that are mostly liquid in nature, are the most frequently encountered. Drinking water, seawater, and some wastewater samples may require only a quick filtration step prior to SPE to remove any sediment that may be present. Beverages and biological fluids may require a more extensive preliminary treatment to remove solids or protein materials that could clog the SPE particles.

Solid samples that are not readily soluble in water, aqueous acid or base, or water–organic solvent mixtures usually require a leaching procedure. Thus, soil samples can be leached by acetone, followed by filtration (1). Many of the constituents of paper and pulp samples can also be recovered by leaching with acetone (2). After dilution with water, conventional SPE can be used.

Gaseous samples include air and atmosphere samples which are frequently analyzed for low concentrations of sulfur dioxide, hydrogen sulfide, nitrogen oxides, organic solvents, hydrocarbons, formaldehyde, and other substances. Often a rather large sample must be used in order to collect sufficient amounts of the impurities for subsequent analysis. Solid-phase extraction by passing the gaseous sample through a cartridge or membrane is usually fast and efficient. Equilibrium of organic components between a gaseous sample and the coated fiber used in SPME is also rapid.

### 9.2.2 Water Samples

Considering the number of published papers (3), this is the largest area by far in which SPE has been used. Each major type of water sample will be illustrated by one or two specific applications, and a number of general references will be given. The extracted substances according to their chemical type are listed in Section 9.5.

***9.2.2.1 Natural Waters*** There has been a great deal of concern over the chemicals that make their way into natural waters (rivers, lakes, etc.) and into supplies of fresh drinking water. In one study agricultural chemicals or their breakdown products were found in over 50% of well-water samples collected in Midwestern United States (4). Numerous studies have been conducted in which SPE was used to concentrate pesticides and other chemicals from various natural water samples (4–17). The need for prefiltration to remove sediment before the SPE step has been stressed (9,14). If this is not done, SPE cartridges and disks may become partially plugged, resulting in low flow rates.

In many of the references cited above, SPE disks loaded with C18 silica or styrene divinylbenzene sorbents instead of cartridges containing loose sorbents were used. However, cartridges filled with graphitized carbon black were used to concentrate neutral and acidic pesticides at low (ng/liter) concentrations (12). The use of liquid–liquid extraction has also been compared to SPE (10–16). Although SPE is generally more efficient, liquid–liquid extraction had the advantage of extracting less of the humic substances from natural water samples (10).

For the most part extracted substances are subsequently eluted from SPE media by a small volume of an organic solvent. However, PAH compounds and some pesticides are sometimes eluted by supercritical carbon dioxide (8,13).

***9.2.2.2 Seawater*** Despite its high salt content, solid-phase extraction of trace organic substances from seawater samples presents no special difficulties. Prefiltration to remove sediment is advised. Specific examples include the SPE of PAH compounds with a glass fiber matrix (18), extraction of phenols using C18 silica cartridges and disks (19), and determination of tributyltin in marine samples (20).

***9.2.2.3 Wastewater*** Analysis of waste effluents is particularly important for several reasons. One is that the level of toxic substances is apt to be considerably higher than that of natural waters. Analysis is essential to monitor and control the amount of toxic substances in the waste stream coming from an industrial plant. It is also important to check on the efficiency of treatment methods used to clean up wastewater. Stricter environmental regulations now require that an industrial plant *prove* that their waste effluents are not unduly polluting.

PAH compounds in wastewater samples have been concentrated by SPE disks (21). In another instance, substances concentrated by SPE were eluted by supercritical carbon dioxide (22). The amounts of surfactants in sewage waste are of concern. Alkane sulfonates in sewage were isolated by SPE after adding a tetrabutylammonium salt to form an ion pair (23). Without this step, the ionic nature of the alkane sulfonate would cause it to be poorly extracted. The evaporated SPE effluent was diluted to 500 μl with isooctane and 2 μl was injected for GC analysis.

In the analytical steps following SPE, it is often sufficient to identify the sample peaks by their known retention times. Quantification is then possible with standard LC or GC techniques. However, more positive identification of chromatographic peaks is often desirable. The method described (24) used liquid chromatography thermospray mass spectrometry (LC-TSP-MS) with selected ion monitoring (SIM) and positive-ion (PI) mode detection.

### 9.2.3 Soil Samples

As a first step it is necessary to extract organic residues from soil samples. In one case (1) this was accomplished by shaking 10 g of soil with acetone to extract PAH residues. The extract was then centrifuged to remove sediment. In another example, sulfonylureas in soil samples were extracted with 0.1 M aqueous sodium hydrogen carbonate (25). After acidification, the extract was subjected to ordinary SPE using C18 silica disks.

### 9.2.4 Agricultural Samples

Many of the applications mentioned in Section 9.2.2 were concerned with analysis of agricultural chemicals that had gotten into various natural waters. However, SPE is only one step in the total analytical process. The following example will provide a more detailed overall perspective.

A study was made of the fungicides captan, captafol, carbendazim chlorothalonil, ethirimol, folpet, metalaxyl, and vinclozolin in environmental waters (24). A 2-liter water sample containing ~2 µg/liter of each fungicide was prefiltered through a 0.45-µm PTFE fiberglass filter to remove particulate matter. A 47-mm C18 silica membrane disk was placed in a vacuum filtration apparatus and washed with 10 ml of methanol and 10 ml of acetonitrile. Without allowing the disk to become dry, 2 liters of water was passed through the disk under vacuum over a period of 1 h. The pesticides taken up by the disk were then eluted with two 10-ml portions of acetonitrile. After careful evaporation of most of the solvent, methanol was added to give a total volume of 500 µl. A 20-µl portion was injected for HPLC analysis.

For analysis by gas chromatography, Durand et al. used a more complex method to extract organophosphorus pesticides and triazines from soil samples (26). Isolation of the pesticide residues was accomplished by Soxhlet extraction of the soil, followed by a cleanup with a Florisil column. The pesticides were finally isolated by SPE using C18 silica Empore disks.

Other examples of pesticides in soil samples with SPE disks include the determination of thiobencarb (27) and seven pesticide residues (atrazine, etc.) (28). In the latter example, the clearest extracts were obtained with 47-mm C8 Empore disks. Basta and Olness determined alachlor, atrazine, and metribuzin in soil by resin extraction (29). Turin and Bowman described a soil extraction method based on SFE for pesticides of varying polarity (30).

### 9.2.5 Food and Beverages

Clear beverages generally are amenable to SPE with little, if any, additional sample pretreatment. SPE of caffeine in soft drinks with C18 silica sorbent is an example (31). In order to determine α- and β-carotene and -cryptoxanthin in orange juice, it was necessary first to centrifuge a 10-ml sample for 5 min (32). This produced a pellet containing the carotenoids. The pellet was extracted several times with 2.0-ml portions of methanol, then the combined extracted carotenoids were saponified with alcoholic potassium hydroxide. After evaporation to dryness and dissolution in methanol, the desired carotenoids finally were isolated by SPE with a C18 silica column.

Identification of synthetic colors in beverage alcohol products has been done by thin-layer chromatography (TLC) following an initial extraction of the colors onto an amino SPE cartridge (33). The SPE step could be done directly on samples of distilled spirits, malt beverages, and wines. However, cream liquors required a centrifuge step to remove the solids.

Analysis of solid food products requires additional pretreatment steps. Thus, powdered samples of mozzarella cheese were hydrolyzed under vacuum with 6 M HCl at 110 °C for 23 h (34). The hydrolysate was derivatized to give a fluorescent derivative of lysinoalanine. This product was then isolated by SPE using an amino extractant cartridge. Natural mozzarella cheese was found to contain 0.4–4 ppm of lysinoalanine where imitation mozzarella cheese contained lysinoalanine levels in the 15–44-ppm range.

Because of its water solubility, determination of organic acids in honey was fairly straightforward (35). The organic acids were taken up by a minicolumn containing a strong-base anion-exchange column. The acids were then eluted by aqueous sulfuric acid.

### 9.2.6 Biological Fluids

Isolation of drugs, drug metabolites, and other relatively small organic substances from biofluids such as urine, whole blood, or plasma is complicated by the presence of proteins or other large molecules in the samples. Protein removal by various precipitating reagents has the disadvantage that some relevant substances may be removed by occlusion. Traditionally, liquid–liquid extraction (LLE) has been used in most clinical and toxicology laboratories. However, SPE is gaining in popularity for removal and isolation of drugs and other substances from biosamples. Compared to LLE, SPE offers the advantages of cleaner extracts, avoidance of emulsion formation, speed, reproducibility, and ease of operation and automation (36,37). Of the numerous publications on SPE for biological fluids, most have dealt with the extraction of a small group of related drugs, and only a few have attempted drug screening (3–6). It is difficult to find a single SPE material capable of efficiently extracting drugs of many classes.

A paper by de Zeeuw (38) emphasized the need for a systematic approach in drug screening. The author outlined an approach called *systematic toxicological analysis* (STA) consisting of three main parts: (1) sample workup—isolation and concentration, (2) differentiation and detection, and

(3) identification. Given the low concentrations of drugs and so on in biofluids and the presence of many interfering substances in the sample matrix, an extraction–concentration step is always required. While SPE is attractive for this step, the requirements of such a method are stringent. The extraction step must be capable of efficiently extracting a very wide variety of substances, ranging from very lipophilic to moderately polar, and extracting acidic, basic, neutral, or zwitterionic compounds.

The scheme proposed by de Zeeuw (38) used mixed-mode SPE columns that exhibit both hydrophobic and cation-exchange interactions. In addition a pH shift was used during the extraction to differentiate between acidic + neutral drugs and basic drugs. The protocol used is outlined in Figure 9.1.

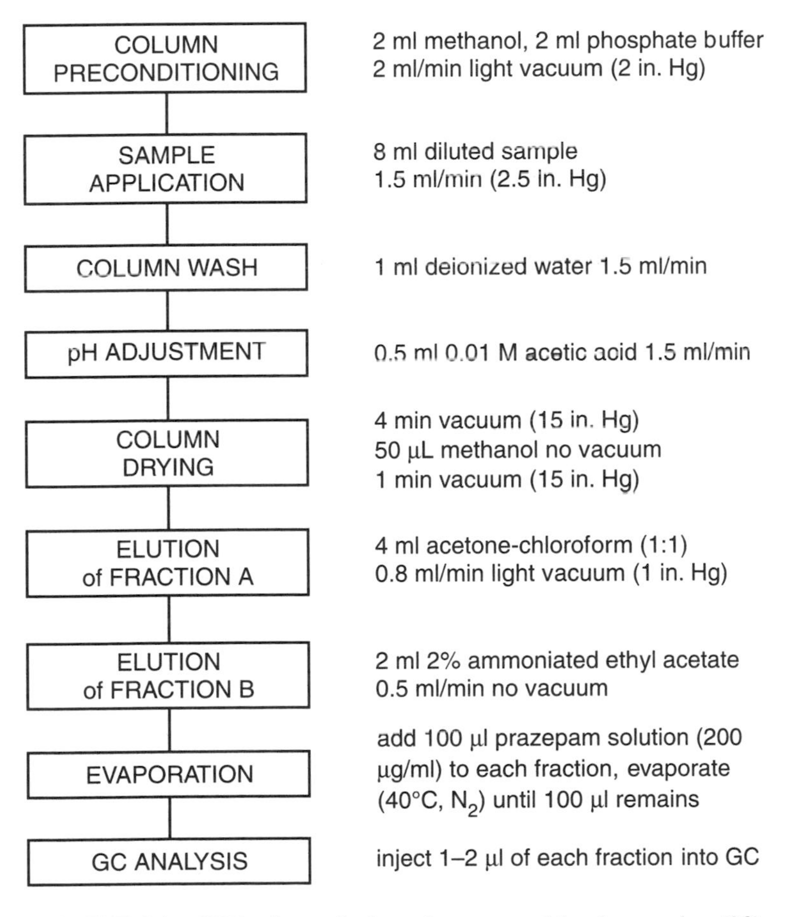

**FIGURE 9.1** SPE scheme for broad-spectrum blood screening (38).

This scheme was found to give satisfactory recoveries for a wide variety of drugs.

Present-day analytical laboratories are concerned more and more with speed and efficiency as well as reliability. A group at the Free University in The Netherlands has devised a system for automated on-line SPE–gas chromatography for determination of benzodiazepines in human plasma (39). The automated system used disposable SPE cartridges and did not require any pretreatment of the plasma samples.

A simple and rugged procedure for determination of amitriptyline, doxepin and their metabolites in porcine serum used an extraction cartridge containing Oasis, which is a relatively hydrophilic extractive resin (40). The serum sample required no preliminary treatment other than the addition of 2% phosphoric acid. SPE isolation of amphetamine and methamphetamine in pharmaceutical and urine samples was possible after first derivatizing the analytes with 1,2-naphthoquinone (41). A C18 silica extraction cartridge was used.

### 9.2.7 Forensics

Chemical analysis is used frequently in resolving issues such as whether illegal drugs have been used or resolving the cause of death in suspect cases. Chemical analysis is also used in crime laboratories. Many of the analytical procedures utilize solid-phase extraction.

Drug screening, discussed in the previous section, also has implications for forensic analysis. In postmortem cases in which cocaine use was suspected, heart, blood, and urine specimens were subjected to SPE prior to analyses by GC-MS (42). Analytes of particular interest included anhydroecgonine methylester, a unique product formed during cocaine smoking, and cocaethylene, which is formed by transesterification of cocaine in the presence of ethanol.

The following are only a few examples of forensic analyses in which SPE plays a vital role:

- Detection of psychotropic drugs in hair (43)—the hair is first dried and powdered prior to work-up for the analysis.
- Determination of the nicotine concentrations in infant hair (44)—taking hair samples rather than body fluids is less painful for infants.
- Arson cases—SPE is used to concentrate accelerants to determine the cause of a fire (45).

### 9.2.8 Paper and Pulp

In the pulp-and-paper industry during the late 1980s, new bleaching processes had to be developed and tested for their ability to deliver pulp-mill effluents that were free of dioxins. Rapid and dependable methods for the toxic dioxins 2378-TCDD and 2378-TCDF were particularly critical. LLE methods for sample preparation were slow and often plagued by emulsions.

A SPE using C18-bonded, 90 mm Empore extraction disks was a key step in solving this analytical problem (46). Two details are worthy of mention. Cleaning the extraction disks by soaking for 1 h in toluene was necessary to reduce the amount of the two dioxins in the disk itself to a very low level, 0.3 pg each. The standards used to calibrate the method were 13 C-labeled dioxins.

A method was needed to estimate newsprint dyes (crystal violet, malachite green, and methylene blue) in effluents (47). The cationic nature of the dyes suggested the use of cation-exchange sorbents in the SPE step. Comparison of sorbents showed that weak-base cation exchangers did not retain the dyes well, but interactions of the dyes with a strong-base cation exchanger were so strong that a suitable elution solvent could not be found. It was finally concluded that a simple visual detection approach was feasible with a strong-base cation exchanger.

Lipid classes in wood and pulp extracts have been determined with the help of SPE (2). Acetone was used to extract the wood and pulp. An aminopropyl silica sorbent was used for SPE.

### 9.2.9 Metals

It is frequently necessary to concentrate metal ions from aqueous samples or to isolate only selected metal ions from complex sample matrices prior to analysis. SPE is an attractive way to do this, particularly if solid sorbents containing selective chelating groups are available. Methods for SPE of metal ions are reviewed in Chapter 7.

In one study a method was needed to separate barium in lake sediments from much larger amounts of alkali- and alkaline-earth metal ions (48). A satisfactory method was devised in which a C18 silica membrane modified with dibenzo-18-crown-6 was used to isolate the barium(II). The major steps in the SPE procedure were as follows: (1) pass 10 mg of dibenzo-18-crown-6 in 2 ml of methanol through a C18 silica Empore membrane, (2) evaporate the solvent and wash with water, (3) add 5 ml of a picric acid solution (picric acid is a good counter anion), and (4) pass a 5-ml sample through the membrane.

Some other examples using SPE to isolate desired metal ions are as follows:

- SPE for concentrating palladium, iridium, and gold (49)
- Concentration of rhodium and palladium β-diketonates (50)
- Determination of lead in soil samples (51)
- SPE of metal ions in vitamins and milk using silica gel containing immobilized 3-hydroxy-2-methyl-l, 4-naphthoquinone (52)

### 9.2.10 Organic Synthesis

In the previous applications SPE has generally been used to concentrate quite small amounts from dilute samples as a prelude to further analysis. However, there is no reason why SPE cannot be used on a larger scale to purify synthetic products by separating them from impurities. Rapid synthesis of drug candidates using combinatorial chemistry is now performed on a very small scale in the pharmaceutical industry, so the use of SPE in organic synthesis need not necessarily be on a large scale. In one instance amides were synthesized and purified by automated SPE (53). The amides were to be used as drug candidates and for rapid generation of structure–activity relationships. The use of SPE resulted in over 90% purity for nearly all compounds synthesized.

SPE has also been used for enrichment of endohedral metallofullerenes (54). Endohedral metallofullerenes are rather polar, while empty fullerenes are nonpolar. Some of the empty fullerenes could be removed without reducing the amount of metallofullerenes.

## 9.3 TECHNIQUES

### 9.3.1 Extraction

Solid-phase extraction of applied samples may be done with cartridges or disks, either off or on line. The SPE mechanism can be one of adsorption, usually by hydrophobic sorbents. Particles with anion or cation-exchange groups may retain analytes by ion exchange as well as by an adsorption mechanism. This dual mechanism for analyte uptake permits group separation of neutral, basic, and acidic analyte types as was described in Chapter 5. The wide variety of solid sorbents used in SPE was discussed in Chapter 3.

Metal ions are retained on SPE particles that contain an appropriate chelating group. Inorganic cations or anions may also be retained by simple ion exchange, although this is usually less selective than chelating resins for metal ions. SPE of metal ions was discussed in Chapter 7.

SPE may also be done by a batch equilibration of the liquid sample with the extractant particles, followed by filtration through an ordinary frit. Batch equilibration is fast if the sorbent particles are small enough (e.g., 5–10 μm). This technique is more tolerant for sediment that might clog a packed SPE column or membrane. The final filtration step provides additional contact of the sample solution with the extractant particles. The amount of solid extractant should be selected so that the sorbent bed after filtration is only 1–2 mm thick. Because the solid particle bed forms gradually during filtration, liquid flow through the bed is more even than it would be using a thin, preformed bed of loose resin. Experiments have showed high recoveries with this technique—as good as with conventional SPE column techniques.

Preserving water samples to prevent deterioration prior to analysis has been a problem. Storage of water samples at 4 °C has been recommended; small amounts of mercuric chloride are sometimes added to inhibit bacterial growth. With increasing numbers of samples being collected, it becomes important to be concerned with storage space as well as storage stability. Field sampling of water samples is readily accomplished by SPE extraction disks, particularly if the disks are in an apparatus where a syringe or hand syringe pump is used to force the sample through the disk. Then the extractive disks can be placed in small protective envelopes and stored until the retained substances can be eluted and analyzed.

In one study the stability of pesticides on Empore SPE disks for 180 days was better than liquid water samples stored for 130 days at 4 °C (16). Another detailed study compared the recoveries of pesticides stored under various conditions (55). Captan gave the most dramatic results. Recovery of captan from water stored for three days at 4 °C was only 28%, whereas the recovery from disks stored at 4 °C was 114%. The author concluded that the stability of pesticides was preserved and in most cases enhanced by concentrating the pesticides on C18 disks. There was some preference for freezing the disks after the SPE extraction, but disks stored at room temperature generally showed good stability on storage.

### 9.3.2 Elution Methods

Extracted organic substances are generally eluted together from the SPE column or disk by a pure organic solvent or mixture of solvents. Sequential elution may be possible by adjusting the acidity or basicity of the eluting solvent so that the chemical forms of some analytes are altered. Sequential elution of neutral and basic fractions or neutral and acidic fractions from ion-exchange sorbents can also be accomplished by elution of the neutral analytes with an organic solvent and then the base or acid fraction with methanol containing methylamine or HCl (see Chapter 5).

Thermal desorption of adsorbed analytes directly into a gas chromatograph is employed routinely in SPME (see Chapter 8). Thermal desorption should also be feasible for conventional SPE by choosing sorbent particles that are stable at >200–250 °C. This approach has already been demonstrated for distinguishing between organic bases and neutral compounds in GC. Placing a sulfonated nonporous polymeric resin in the injection port sleeve of the GC retained the bases as protonated cations while allowing neutral species to pass unchanged into the GC column (56).

Extracted analytes may also be eluted from SPE disks by a supercritical fluid (SF). Use of a SF has some advantages over liquid organic solvents for this purpose. Supercritical fluids such as carbon dioxide are not toxic, noncombustible, chemically inert and easy to discard. SF carbon dioxide acts as a hydrophobic solvent. It may be necessary to add a modifier such as methanol for efficient elution of less hydrophobic analytes. Elution of phenols from SPE disks (57), and organochlorine and organophosphorous pesticides from Empore disks (58) are typical examples.

### 9.3.3 Methods Not Requiring Elution

Membrane disks used for SPE are both thin (≤1 mm) and efficient. Tests with extractable dyes show that analytes tend to be evenly adsorbed in a thin layer on the upper surface of the membrane. These properties make it feasible to measure adsorbed analytes directly on the membrane itself. This completely eliminates the need for a solvent elution step.

A method for $^{90}Sr^{2+}$ uses Empore Rad Disks (3M Co.), which are impregnated with particles of a chelating polymer, AnaLig (from IBC Advanced Technologies, Inc.) when a liquid sample is passed through the disk, $^{90}Sr^{2+}$ is selectively adsorbed in a shallow surface layer on the membrane. The disk is then dried and placed into an appropriate counter to measure the amount of radioactive $^{90}Sr^{2+}$ on the disk. This method is fast and accurate with low reagent and reagent preparation costs.

Amphetamines have been derivatized on a SPE support and the colored product measured photometrically (41).

After extraction onto a SPE membrane, pesticides can be determined by matrix-assisted laser desorption/ionization–mass spectrometry (MALDI-

MS) (59). The SPE membrane is placed on the target where pulses from a laser ablate the membrane surface and ionize the analytes. A repeller introduces the ions into an ion lense for measurement. Quantitative determinations required the use of an internal standard.

Other methods can of course be used to measure analytes deposited on a surface. For example, in a technique related to SPE, disposable IR cards are used for infrared measurements within the spectral range of 4000 to 400 $cm^{-1}$ (60). Each card has a 19-mm circular aperature containing a thin, microporous film for sample application. Liquid samples can be applied to the film in amounts up to 50 μl. The microporous film enhances solvent evaporation under ambient conditions. Infrared measurements can then be made conveniently on the sample film.

## 9.4 PROBLEMS WITH PRACTICAL SAMPLES

Application of SPE to particle-laden water samples can be a vexing problem. The analytes of interest may be adsorbed onto the particulates as well as dissolved in the liquid sample. When this occurs it may be necessary to filter off the particulate matter and extract the adsorbed materials with an appropriate solvent. Strongly adsorbed substances may necessitate repeated extraction with a liquid solvent or a supercritical fluid. Another concern is that suspended particles may plug the SPE cartridge or membrane and decrease the flow rate so much that it effectively stops. The worst plugging problems are most often encountered with surface waters that are high in biological activity or with waters containing fine suspended clay particles. Some hard waters may contain a very fine suspension of calcium carbonate.

The severity of plugging depends on the type, concentration, and size of the particulates in the sample; the pore size of the sorbent; and the area of the sorbent bed. The smaller size of particles used in SPE disks (~10 μm) renders disks generally somewhat more susceptible to plugging than cartridges, which customarily contain particles of around 40 μm.

Larger particulate matter can be removed before the SPE step by a prefilter that is a mat of distinct pore size that filters particulates mechanically. For samples that contain particulates too fine to be removed effectively by a prefilter it is best to use a filtration aid. The function of a filtration aid is to adsorb particulate matter from the sample and prevent it from clogging the pores of the SPE disk or cartridge. The filtration aid may be placed on top of the disk as a thin layer or it may be added to the sample to form a layer during the SPE step.

**TABLE 9.1 Time (min) Needed to Pass 1 Liter of Aqueous Samples Through a 90-mm SPE Disk**

| Sample Type | SPE Disk Only | SPE Disk Glass Fiber Filter | SPE Disk Glass Beads |
|---|---|---|---|
| Lake water | 20 | 11 | 9 |
| River water | 40 | 9 | 4 |
| Pulp and paper | 19 | — | 8 |
| Stagnant pond | 985 | 572 | 221 |

*Source:* Data from Reference 62.

Diatomaceous earth, perlite, silica granules, and other substances have been used as filter aids. However, one detailed study found glass beads to be an exceptionally good filter aid for use with SPE disks (61). The beads were made of a high-density metal oxide glass with an average particle size of 40 μm. They are now available commercially (3M Co.). The effectiveness of glass beads as a filter aid was determined by measuring the time required to pass a 1-liter sample through a 90-mm SPE disk under fixed conditions (see Table 9.1).

Samples that contain proteins or other large biomolecules can pose a serious problem. These biomolecules stick easily to surfaces and can readily plug up SPE cartridges or disks. Several methods can be employed to

**TABLE 9.2 SPE Applications According to Analyte Class**

| Analyte Class | References |
|---|---|
| Aromatic sulfonates | 7, 65–67 |
| Drugs | 38–41, 43, 44, 69–73 |
| Dyes | 33, 47, 75, 76 |
| Explosives | 11, 71, 72 |
| Herbicides and fungicides | 1, 24, 25, 73, 74 |
| Metal ions | 49–53, 48, 50, 51, 75, 76 |
| Oil and grease | 77–80 |
| Organometallics | 20, 54 |
| PAH compounds | 21, 81–83 |
| Pesticides | 8, 10, 12–17, 22, 26–30, 41, 58, 84, 85–88 |
| Synthetic organics | 22, 89 |
| Miscellaneous organics | 19, 32, 35, 46, 58, 81, 93–100 |

minimize these difficulties. Oasis HLB extraction cartridges (Waters Co.) contain a polystyrene–vinylpyrrolidone copolymer that has a moderately hydrophilic surface. It has been successful in extracting many types of drugs and other small molecules directly from serum and urine samples. A unique type of molecular sieve known as *Silicalite* can also extract many smaller molecules directly from biosamples (62). The surface of Silicalite is hydrophilic, and adsorption occurs mainly within pores that are small enough to reject large molecules. A considerable number of sorbents with restricted access have been used to exclude large molecules in direct-injection HPLC. These should also be useful for SPE. Boos and Rudolphi have written a comprehensive review on restricted-access media (63).

## 9.5 APPLICATIONS ACCORDING TO CHEMICAL CLASS

This section gives typical references to the use of SPE on "real" samples, classified according to the chemical type of the analytes. The examples in Table 9.2 illustrate the practical applications of SPE; no attempt has been made to provide an exhaustive list.

## REFERENCES

1. P. R. Koostra, M. H. C. Straub, G. H. Stil, E. G. van der Velde, W. Hesselink, and C. C. J. Land, *J. Chromatogr. A* **697** (1995) 123–129.
2. T. Chen, C. Breuil, and S. Carriere, *TAPPI J.* **77** (1994) 235–240.
3. *AGRICOLA Index* (1985–1996), *CCAB Index* (1996–1997), *ASTI Index* (1983–1996).
4. D. W. Kolpin, *J. Env. Qual.* **26** (1997) 1025–1037.
5. A. D. Corcia, C. Crescenzi, and E. Guerriero, *Env. Sci. Technol.* **31** (1997) 1658–1663.
6. S. Chiron, A. Valerde, A. A. Fernandez, D. Barceló, *J. AOAC* **78** (1995) 1346–1352.
7. Y. Yamini and M. Ashraf-Khorassani. *Fresenius. J. Anal. Chem.* **348** (1994) 251–252.
8. P. H. Tang, J. S. Ho, and J.W. Eichclberger, *J. AOAC Int.* **76** (1993) 72.
9. T. M. McDonnel and J. Rosenfeld, *J. Chromatogr.* **629** (1993) 41–53.
10. S. Chiron, A. F. Alba, and D. Barceló, *Env. Sci. Technol.* **27** (1993) 2352.
11. G. LeBrun, P. Rethwill, and J. Matterson, *Env. Lab.* (Feb/March 1993).
12. T. D. Bucheli, F. C. Gruebler, S. R. Miller, and R. P. Schwarzenbach, *Anal. Chem.* **69** (1997) 1569.
13. J. S. Ho and W. L. Bodde, *Anal. Chem.* **66** (1994) 3716.

14. S. Bengtsson, T. Berglöf, S. Granat, and G. Jonsäll, *Pest. Sci.* **41** (1994) 55.
15. G. Molina, M. Honing, and D. Barceló, *Anal. Chem.* **66** (1994) 4444.
16. D. Barceló, S. Chiron, S. Lacorte, E. Martinez and J. S. Salau, *Trends Anal. Chem.* **13** (1994) 352.
17. B. A. Tomkins, R. Merriweather, and R. A. Jenkins, *J. AOAC Int.* **75** (1992) 1091.
18. I. Urbe and J. Ruana, *J. Chromatogr. A* **778** (1997) 337–345.
19. M. T. Galceron and O. Jáuregui, *Anal. Chim. Acta* **304** (1995) 75.
20. O. Evans, B. J. Jacobs, and A. L. Cohen, *Analyst* **116** (1991) 15.
21. D. Eastwood, M. E. Dominguez, R. L. Lidberg, and E. J. Poziomek, *Analysis* **22** (1994) 305.
22. V. S. Ong and R. A. Hites, *Env. Sci. Technol.* **29** (1996) 1259.
23. J. A. Field, T. M. Field, T. Polger, and W. Giger, *Env. Sci. Technol.* **28** (1994) 497.
24. J. S. Salau, R. Alonso, G. Batlló, and D. Barceló, *Anal. Chim. Acta* **293** (1994) 109.
25. P. Klaffenbach and P. T. Holland, *J. Agric. Food Chem.* **41** (1993) 396.
26. G. Durand, R. Alonso, and D. Barceló, *Quimica Analitica* **10** (1991) 157.
27. M. J. Redondo, M. J. Ruiz, R. Boluda, and G. Font, *J. Chromatogr. A* **678** (1994) 375.
28. M. J. Redondo, M. J. Ruiz, R. Boluda, and G. Font, *Chromatographia* **36** (1993) 187.
29. N. T. Basta and A. Olness, *J. Env. Qual.* **21** (1992) 497.
30. H. J. Turin and R. S. Bownan, *J. Env. Qual.* **22** (1993) 332.
31. Y. Daghbouche, S. Garrigues, and M. T. Vidal, *Anal. Chem.* **69** (1997) 1086–1091.
32. J. F. Fisher and R. L. Rouseff, *J. Agric. Food Chem.* **34** (1986) 985.
33. S. M. Dugar, J. N. Leibowitz and R. H. Dyer, *J. AOAC Int.* **77** (1994) 1335.
34. L. Pellegrino, P. Resmini, I. De Noni, and F. Masotti, *J. Dairy Sci.* **79** (1996) 725–734.
35. A. Cherchi, L. Spanedda, C. Tuberoso, and P. Cabras, *J. Chromatogr. A* **669** (1994) 59–64.
36. R. D. McDowall, *J. Chromaotogr.* **492** (1989) 3.
37. R. E. Majors, *LC-GC Int.* **4** (1991) 10.
38. R. A. de Zeeuw, *J. Chromatogr. B* **689** (1997) 71.
39. A. J. H. Louter, E. Bosma, J. C. A. Schipperen, J. J. Vreuls, and U. A. Th. Brinkman, *J. Chromatogr. B* **689** (1997) 35.
40. Y. F. Cheng, D. J. Phillas, E. Neve, and L. Bean, *J. Liq. Chrom. Rel. Technol.* **20** (1997) 2461.
41. P. C. Falcó, C. M. Legua, A. S. Cabeza, and R. Porras, *Analyst* **122** (1997) 673.
42. A. J. Jenkins and B. A. Golberger, *J. Forensic Sci.* **42** (1997) 824–827.
43. M. Yegles, F. Mersch, and R. Wennig, *Forensic Sci. Int.* **84** (1997) 211–218.

44. S. Pichini, I. Altieri, M. Pellegrini, R. Pacifici, and P. Zuccaro, *Forensic Sci. Int.* **84** (1997) 253–258.
45. W. Bertsch, *Anal. Chem.* **68** (1996) 541A–545A.
46. S. Barkowski, *Env. Lab.* **12** (March/April 1995).
47. E. Milanova and B. B. Sithole, *TAPPI J.* **80** (1997) 121–128.
48. Y. Yamini, N. Alizadeh, and M. Shamsipur, *Sep. Sci. Technol.* **32** (1997) 2077–2085.
49. Y. A. Zolotov, *J. Anal. Chem.* **51** (1996) 941.
50. B. W. Wenclawawiak, T. Hees, C. E. Zöller, and H. P. Kabus, *Fresenius J. Anal. Chem.* **358** (1997) 471.
51. P. Sooksamiti, H. Geckeis, and K. Grudpan, *Analyst* **121** (1996) 1413.
52. B. S. Garg, J. S. Bist, R. K. Sharma, and N. Bhojak, *Talanta* **43** (1996) 2093.
53. R. M. Lawrence, S. A. Biller, O. M. Fryszman, and M. A. Poss, *Synthesis-Stuttgart* **5** (1997) 553.
54. H. W. Zhang, K. P. Kwong, T. Wong, H. Shinohara, and M. Inakuma, *Tetrahedron Lett.* **37** (1996) 9127–9130.
55. S. A. Senseman, *Env. Lab.* (Oct./Nov. 1992).
56. J. J. Sun and J. S. Fritz, *J. High Resol. Chromatogr.* **14** (1991) 69.
57. P. H. Tang, *J. High Resol. Chromatogr.* **17** (1994) 509.
58. I. J. Barnabas, J. R. Dean, S. M. Hitchen, and S. P. Owen, *Anal. Chim. Acta* **291** (1994) 261.
59. A. W. T. Bristow, C. S. Creaser, S. Nelieu, and J. Einhorn, *Analyst* **121** (1996) 1425–1428.
60. 3M Company, *Disposable IR Cards*, application notc A-1.
61. T. A. Dirksen, S. M. Price, and S. J. St. Mary, *Am. Lab.* (Dec. 1993).
62. D. Ambrose and J. S. Fritz, *J. Chromatogr. A* **773** (1997) 189; **771** (1997) 45.
63. K. S. Boos and Anne Rudolphi, *LC-GC* **15** (1997) 602, 814.
64. B. Altenbach and W. Giger, *Anal. Chem.* **67** (1995) 2325–2333.
65. M. Frost, H. Kohler, and G. Blaschke, *Int. J. Legal Med.* **109** (1996) 53.
66. D. A. McLoughlin, T. V. Olah, and J. D. Gilbert, *J. Pharmacol. B* **15** (1997) 1893.
67. A. J. Tomlinson, L. M. Benson, S. Jameson, D. H. Johnson, and S. Naylor, *J. Am. Soc. M* **8** (1997) 15–24.
68. S. Pálmarsdóttir, E. Thordarson, L.-E. Edholm, J. Å. Jonsson, and L. Mathiasson, *Anal. Chem.* **69** (1997) 1732–1737.
69. M. J. I. Mattina and G. J. MacEachern, *J. Chromatogr. A* **679** (1994) 269.
70. A. J. Borgerding and R. A. Hites, *Env. Sci. Technol.* **28** (1994) 1278.
71. T. F. Jenkins, P. H. Miyares, K. F. Myers, E. F. McCormick, and A. B. Strong, U.S. Army Corps Engineers, Special Report, Dec. 1992.
72. EPA Method 3535A.
73. J. A. Field and K. Monohan, *J. Chromatogr. A* **741** (1993) 85.
74. A. Balinova, *J. Chromatogr. A* **728** (1996) 319.
75. K. Ohto, Y. Tanaka, and K. Inoue, *Chem. Lett.* **7** (1997) 647.

76. G. Malofeeva, O. Petrukhin, L. Rozhkova, B. Spivakov, G. Genkina, and T. Mastryukova, *Russ. J. Anal. Chem.* **51** (1996) 1061.
77. C. K. Cross, *J. Am. Oil Chemists Soc.* **67** (1990) 142.
78. S.-L. Lau, and M. Stenstrom, *Water Env. Res.* **69** (1997) 368.
79. R.-G. Marquez, N. Jorge, M. Martin-Polvillo, and M. Dobarganes, *J. Chromatogr. A* **749** (1996) 55–60.
80. EPA Method 1664.
81. L. Rivera, M. J. C. Curto, P. Pais, M. T. Galceran, and L. Puignou, *J. Chromatogra. A* **731** (1996) 85.
82. P. R. Loconto, *J. Chromatogr. A* **774** (1997) 223.
83. EPA Method 550.1.
84. F. J. Schenick, R. Wagner, M. K. Hennessey, and J. L. Okrasinski, Jr., *J. AOAC Int.* **77** (1992) 1036.
85. W. C. Quayle, I. Jepson, and I. A. Fowlis, *J. Chromatogr. A* **773** (1997) 271.
86. D. Barceló, G. Durand, V. Bouvot, and M. Nielen, *Env. Sci. Technol*, **27** (1993) 271.
87. J. Patsias and M.-E. Papadopoulou-Mourkidou, *J. Chromatogr. A* **740** (1996) 83.
88. L. M. Davi, M. Baldi, L. Penazzi, and M. Liboni, *Pest. Sci.* **35** (1992) 63–67.
89. C. Vogel, *Synthesis-Stuttgart* **5** (1997) 497ff.
90. T. D. Power and J. F. Sebastian, *Tetrahedron Lett.* **37** (1996) 912.
91. Y. C. Ling, M. Y. Chang, and I. P. Huang, *J. Chromatogr. A* **669** (1993) 119.
92. J. W. Pensabene, W. Fiddler, and R. A. Gates, *J. AOAC Int.* **75** (1992) 438.
93. D. Puig, I. Silgoner, M. Grasserbauer, and D. Barceló, *Anal. Chem.* **69** (1997) 2756.
94. N. Masque, M. Galia, R. M. Marcé, and F. Borrull, *J. Chromatogr. A* **771** (1997) 55.
95. K. Z. Taylor, D. S. Waddell, E. J. Reiner, and K. A. MacPherson, *Anal. Chem.* **67** (1995) 1186.
96. EPA Method 525.1.
97. EPA Method 506.
98. EPA Method 1613.
99. EPA Method 552.1.
100. EPA Method 515.2.

# INDEX